Get the most from this book

This revision guide has been written to accompany the Edexcel GCSE (9–1) Geography B specification to help you get the best possible result in your examinations.

This books aims to give you the essentials that should serve as a reminder of what you will have covered in your course and allow you to bring together your own learning and understanding.

Everyone has to decide his or her own revision strategy, but it is essential to review your work, learn it and test your understanding. These revision notes will help you to do that in a planned way, topic by topic. Use this book as the cornerstone of your revision and don't hesitate to write in it – personalise your notes and check your progress by ticking off each section as you revise.

Tick to track your progress

Use the revision planner on pages 4–7 to plan your revision, topic by topic. Tick each box when you have:

- revised and understood a topic
- tested yourself
- practised the exam questions and gone online to check your answers.

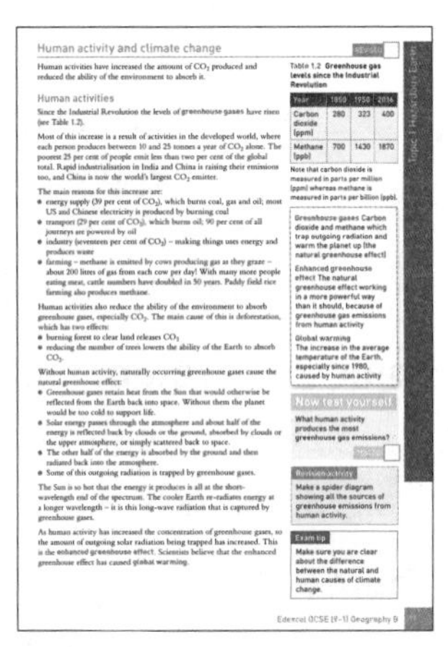

You can also keep track of your revision by ticking off each topic heading in the book. You may find it helpful to add your own notes as you work through each topic.

Features to help you succeed

This guide contains features intended to help you to *actively* work through your revision schedule.

Revision activities

These activities have been designed to focus your revision. If you can complete these activities, congratulate yourself and give yourself a reward. If you find some of the activities challenging, read over your notes again and speak to your teacher if you need further help.

Case studies and located examples

It is vital that you can give specific examples, particularly in the questions requiring extended answers. This demonstrates the detailed knowledge examiners are looking for to award the higher grades. Often, your example does not require lots of detail but should be used to back up a point being made rather than simply recounted out of context.

Now test yourself

These activities are designed to help you decide if you fully understand a topic and are exam ready. If you struggle, then read over your notes again and perhaps ask to see your teacher or bring the topic up in revision classes.

Exam practice

These questions closely resemble the style of the question that you will face in the examination. Use them to consolidate your revision and practise your exam skills.

Exam tips

Expert tips are given throughout the book, including identifying common mistakes and suggesting strategies for getting the best out of the time you have in the examination room. These will therefore help to boost your final grade.

Definitions and key terms

Here you will find some of the specialist geography terminology that you need to know. Key terms are highlighted in bold throughout the book. Clear, concise definitions are provided where the essential key terms first appear.

Online

Go online to check your answers to the now test yourself questions and the exam practice questions at **www.hoddereducation.co.uk/myrevisionnotes**

My revision planner

Component 1 Global Geographical Issues

Component 2 UK Geographical Issues

Component 3 People and Environment Issues

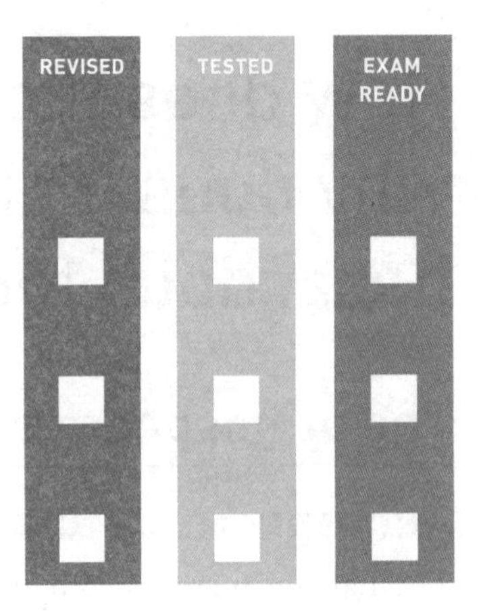

Exam advice

You have three GCSE Geography exams. All are 1 hour 30 minutes long:

- Paper 1 Global Geographical Issues: topics 1–3
- Paper 2 UK Geographical Issues: topics 4–6
- Paper 3 People and Environment Issues – Making Geographical Decisions: topics 7–9.

The exams will have questions that range from 1 mark up to 16 marks. Exam questions use different command words. These words tell you what type of answer is expected. The three most often used command words are shown below:

Describe Usually a 2-mark question	Describe means 'what' or 'say what you see' Usually a 'describe' question involves looking at a figure, such as a graph, map or image Use specific data from the figure to describe a pattern, distribution, trend or change If there is number data on the figure, use it in your answer
Explain Usually a 2- or 4-mark question	Explain means 'why' You have to give reasons for a pattern or process Try to use the word 'because' in your answer Give a basic explanation, then add more detail to it like this: *'In developing countries, fertility rates are high* ***because*** *children work from a young age providing family income, but as family incomes rise this is needed less so women have fewer children.'* Extended explanations like this will score 2 marks
Assess Usually an 8- or 12-mark question	Assess means weighing-up the importance of different factors, and deciding which are the most important An 'assess' question needs a conclusion, where you make a judgement about which factor is the most important You also need to use data and examples as evidence to back up your points and judgement Try to use words like 'however' and 'but' and 'on the other hand' to show you are weighing things up

Topic 1 Hazardous Earth

How does the world's climate system function, why does it change and how can this be hazardous for people?

The atmosphere as a global system

REVISED

Atmospheric circulation

The Earth is surrounded by a thin layer of air called the atmosphere, which moves in response to differences in temperature at the equator (warm) and the poles (cold). The movement of air is called **atmospheric circulation**. It happens because incoming solar radiation (sunlight) strikes the equator at almost 90° to the surface, causing intense heating. Towards the poles **solar radiation** strikes the surface at a much lower angle and the heat is spread over a much greater surface area, leading to colder temperatures.

Heat from the equator is moved towards the poles in two ways, by:

- atmospheric circulation cells: the Hadley, Ferrel and Polar cells
- ocean currents.

The overall effect of this heat transfer is to even out global temperature between the poles and equator by redistributing heat energy.

Atmospheric circulation The pattern of air flow in the atmosphere, dominated by three large circulation cells in each hemisphere

Solar radiation Heat energy from the Sun received at the Earth's surface

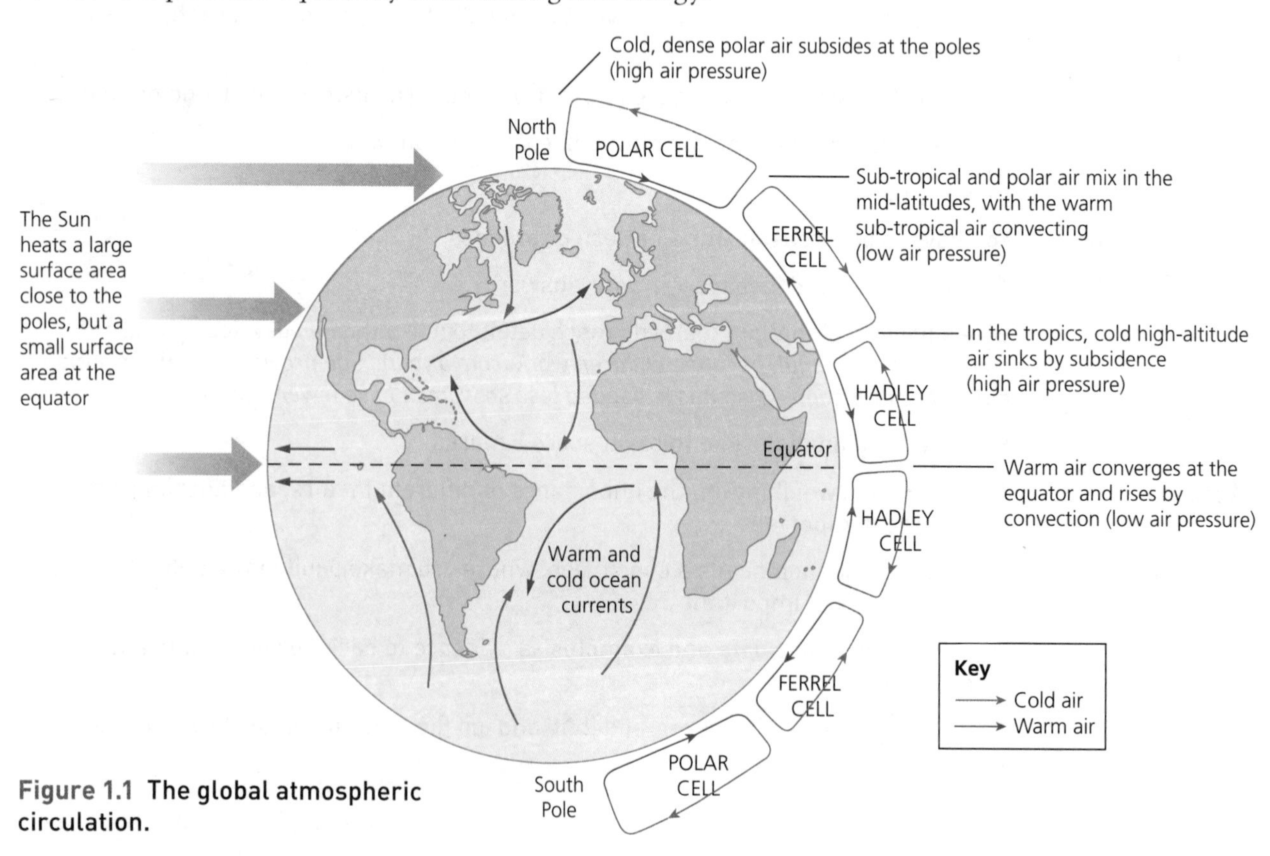

Figure 1.1 The global atmospheric circulation.

The average temperature at the equator is about 30 °C whereas at the poles it is about –30 °C. This difference would be greater if **heat transfer** didn't take place. Ocean currents act like 'rivers' of water in the sea and can be warm or cold. Cold ones move towards the equator, whereas warm ones move towards the poles.

Heat transfer The movement of heat from the equator towards the poles

Latitude The position on the Earth's surface, in degrees, north or south from the equator

Now test yourself

TESTED

How many atmospheric cells are in each hemisphere?

Arid and high-rainfall areas

The three atmospheric circulation cells shown in Figure 1.1 encircle the entire planet, like vast rotating 'tubes'. They are mirrored in the northern and southern hemispheres. Air moves in the atmosphere either towards the ground (subsidence) or up into the atmosphere (convection). These movements have a big influence on air pressure and rainfall (see Table 1.1).

Revision activity

It's useful to be able to sketch a diagram of Earth's atmospheric cells, so practise doing this and labelling it.

Table 1.1 Air movement causes, air pressure and rainfall

Air movement	Cause	Air pressure	Cloud and rain
Convection	High amounts of solar radiation heat the ground, which heats the air above the ground The heated air rises As the air rises it cools and condenses, forming water droplets and clouds	Low air pressure, because air is moving upwards away from the surface	Thick cloud and heavy rainfall
Subsidence	In places with very low-intensity solar radiation like the poles, or where air is very cold at high altitude, the cold, dense air sinks towards the ground As the air sinks it warms up, so can hold more moisture and this prevents clouds from forming	High air pressure, because air is 'piling down' towards the surface	No, or very thin, cloud and very little rainfall

The atmospheric circulation cells encircle the Earth in horizontal bands, creating a broad pattern of low-pressure (high rainfall) and high-pressure (low rainfall) belts distributed in bands by **latitude** (see Figure 1.2).

The distribution of oceans and continents affects the location of high- and low-pressure areas, so the real pattern is not as symmetrical as expected.

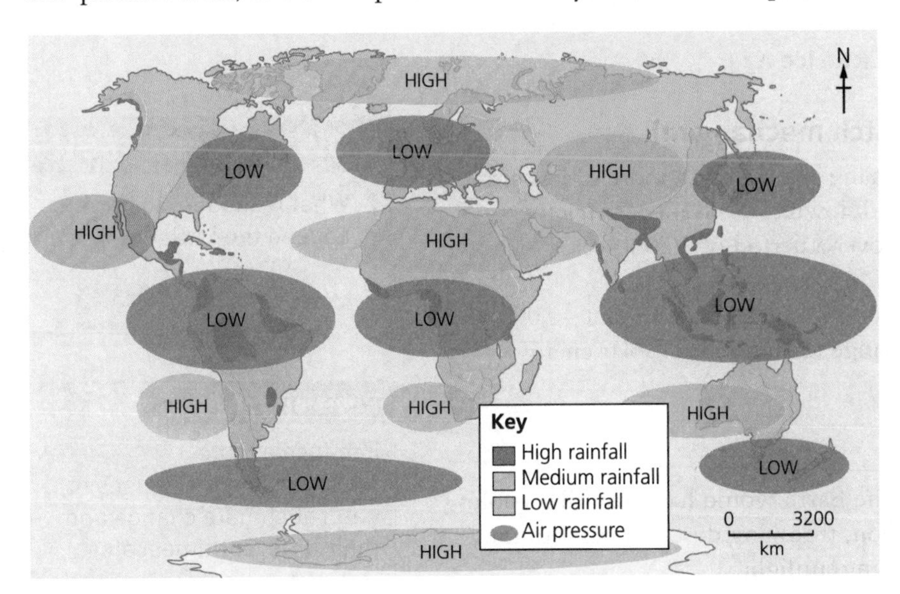

Figure 1.2 Annual rainfall and high- and low-pressure areas.

Now test yourself

TESTED

1 How do temperature and rainfall change as you move away from the equator towards the North and South Poles?

Exam practice

1 Describe the location of high-pressure areas in Figure 1.2 (page 7). [2]
2 Explain two ways in which heat is transferred from the equator towards the poles. [4]

ONLINE

Exam tip

If you are asked to 'describe a pattern' in the exam, start with an overview of the main pattern, rather than starting with the detail.

The natural causes of climate change

REVISED

There are several theories about past **climate change**. It is possible that these processes operate together, in which case climate change would be quite severe, or perhaps they 'pull' in opposite directions, in which case the changes would not be as large.

Climate change Any change (short or long term) in the Earth's average temperature and/or precipitation patterns

Solar output The amount of solar radiation being emitted by the Sun

Little Ice Age A period from about 1300 to 1870 when average temperatures were at least 1 °C cooler than the present time

Orbital eccentricity Change in the shape of the Earth's orbit around the Sun

The main causes

The four most important causes are:

1 Volcanic activity

- Large eruptions emit vast quantities of dust and gases such as sulphur dioxide into the atmosphere.
- The dust and gases block out or absorb incoming solar radiation, so the Earth cools.
- Examples include Mt Pinatubo in 1991, the Laki eruption in 1783 and Mt Toba 70,000 years ago.

2 Variations in solar output

- Sunspots are darker areas on the Sun's surface – they are a sign of greater **solar output**.
- Sunspots come and go in cycles of about eleven years.
- However, there are longer periods when very few sunspots were observed, such as 1645–1715.
- This period coincides with the **Little Ice Age**.

3 Orbital changes (Milankovitch mechanism)

- The shape of the Earth's orbit changes (becoming more or less circular) over a period of 100,000 years – known as **orbital eccentricity**.
- The Earth 'wobbles' on its axis over a period of 26,000 years – known as precession.
- The tilt of the axis varies between 21° and 24° over about 40,000 years.
- Taken together, these effects change the amount of solar energy received at the Earth's surface.

4 Asteroid collisions

- A large asteroid colliding with the Earth would have a similar cooling effect to a major volcanic eruption, throwing dust and ash into the atmosphere and blocking incoming sunlight.

Now test yourself

2 What is thought to have caused the Little Ice Age?

TESTED

Revision activity

Produce a table summarising the causes of natural climate change and their different timescales.

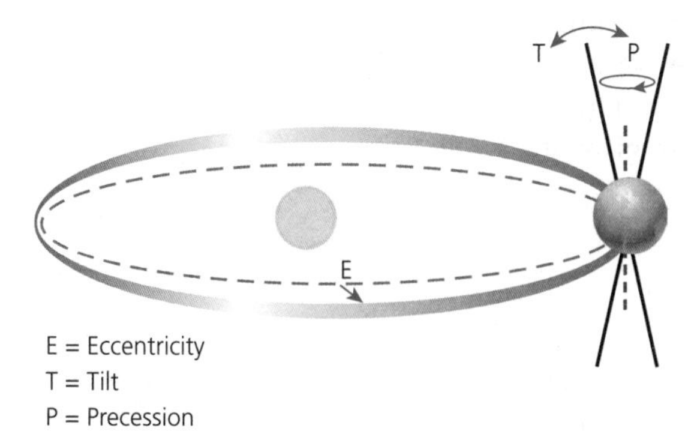

Figure 1.3 The Milankovitch mechanism.

All causes of climate change can operate together, making the planet hotter or colder, but they might also cancel each other out. There is a lot of evidence that the oceans play a key role in controlling global temperatures and if this gets out of balance then change can accelerate.

Evidence for climate change

The weather can change from minute to minute; in the UK it often does! Climate is defined as the average weather conditions over 30 years.

We know that climate has changed a great deal in the past. This is shown by:

- Evidence from **ice cores** in Greenland and Antarctica showing how much carbon dioxide (CO_2) was in the atmosphere when the ice was formed.
- Evidence from tree rings; annual growth rings in trees preserve a record of that year's growth conditions, so a wide ring means it was warm whereas a narrow ring indicates harsh conditions like cold or drought.
- Evidence from historical sources like paintings, diaries and harvest records – although these tend to be unreliable and incomplete.

Ice cores Tubes of ice drilled out from ice sheets; they contain trapped air bubbles and the air dates from when the ice was formed

Quaternary The most recent geological time period that began 2.6 million years ago

Figure 1.4 shows a regular pattern of high and low temperatures, like a cycle. This period, known as the **Quaternary**, consisted of cold glacial periods with ice sheets growing to cover much of Europe, Asia and North America. There were also shorter interglacial periods which were warmer with a climate much like today's.

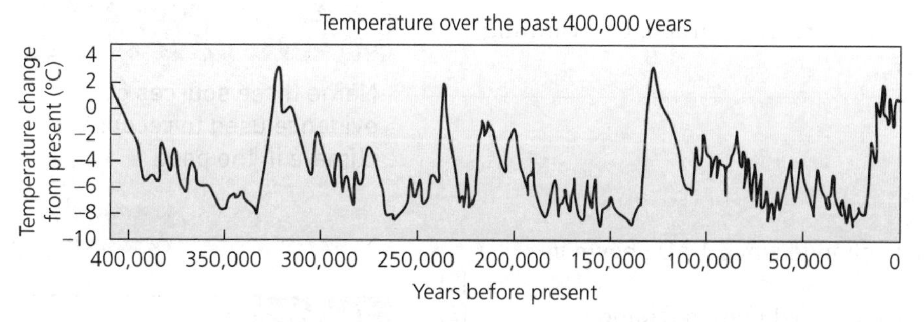

Figure 1.4 Temperature change over the past 400,000 years of the Quaternary period.

In historical times, over the last few thousand years (since Roman times), temperatures have varied by as much as 1.5 °C each side of the average. It only takes very small changes in average temperatures to make a great deal of difference to what farmers can grow and where they can grow it.

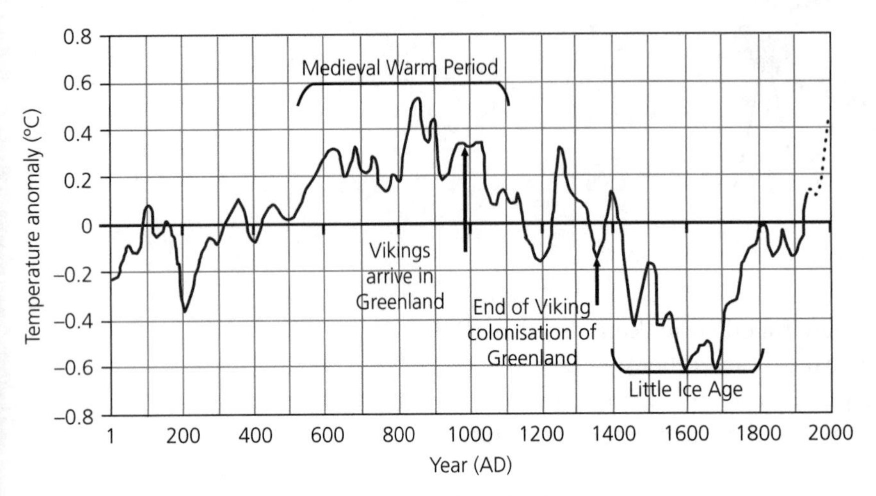

Figure 1.5 Temperature changes in historical times.

Case study: the Little Ice Age

One of the best known periods of climate change in the recent past is the Little Ice Age, probably caused by reduced sunspot activity. It lasted from about 1300 to 1870 (see Figure 1.5). Average temperatures were 1 °C below today's.

Impacts of the Little Ice Age included:

- The Baltic Sea froze over in winter, as did rivers in Europe including the Thames.
- Sea ice, which today is far to the north, reached as far south as Iceland.
- Winters were much colder and longer, reducing the growing season by several weeks.
- These conditions led to widespread crop failure and famine.
- Remote areas such as Greenland were abandoned by settlers as survival became impossible.
- The price of grain increased almost everywhere, leading to social unrest and revolt.
- Glaciers advanced in the Alps and northern Europe, overrunning towns and farms in the process.

Exam practice

1. Describe the pattern of climate change since 1 AD shown in Figure 1.5. [2]
2. Explain two sources of evidence for past climate change. [4]
3. Assess the extent to which the Earth's climate has changed in the past due to natural causes. [8]

ONLINE

Now test yourself

Name three sources of evidence used to reconstruct climate in the past.

TESTED

Exam tip

Learn some key facts about case studies to add detail to your longer answers.

Human activity and climate change

REVISED

Human activities have increased the amount of CO_2 produced and reduced the ability of the environment to absorb it.

Human activities

Since the Industrial Revolution the levels of **greenhouse gases** have risen (see Table 1.2).

Most of this increase is a result of activities in the developed world, where each person produces between 10 and 25 tonnes a year of CO_2 alone. The poorest 25 per cent of people emit less than two per cent of the global total. Rapid industrialisation in India and China is raising their emissions too, and China is now the world's largest CO_2 emitter.

The main reasons for this increase are:

- energy supply (39 per cent of CO_2), which burns coal, gas and oil; most US and Chinese electricity is produced by burning coal
- transport (29 per cent of CO_2), which burns oil; 90 per cent of all journeys are powered by oil
- industry (seventeen per cent of CO_2) – making things uses energy and produces waste
- farming – methane is emitted by cows producing gas as they graze – about 200 litres of gas from each cow per day! With many more people eating meat, cattle numbers have doubled in 50 years. Paddy field rice farming also produces methane.

Human activities also reduce the ability of the environment to absorb greenhouse gases, especially CO_2. The main cause of this is deforestation, which has two effects:

- burning forest to clear land releases CO_2
- reducing the number of trees lowers the ability of the Earth to absorb CO_2.

Without human activity, naturally occurring greenhouse gases cause the natural greenhouse effect:

- Greenhouse gases retain heat from the Sun that would otherwise be reflected from the Earth back into space. Without them the planet would be too cold to support life.
- Solar energy passes through the atmosphere and about half of the energy is reflected back by clouds or the ground, absorbed by clouds or the upper atmosphere, or simply scattered back to space.
- The other half of the energy is absorbed by the ground and then radiated back into the atmosphere.
- Some of this outgoing radiation is trapped by greenhouse gases.

The Sun is so hot that the energy it produces is all at the short-wavelength end of the spectrum. The cooler Earth re-radiates energy at a longer wavelength – it is this long-wave radiation that is captured by greenhouse gases.

As human activity has increased the concentration of greenhouse gases, so the amount of outgoing solar radiation being trapped has increased. This is the **enhanced greenhouse effect**. Scientists believe that the enhanced greenhouse effect has caused **global warming**.

Table 1.2 Greenhouse gas levels since the Industrial Revolution

Year	1850	1950	2016
Carbon dioxide (ppm)	280	323	400
Methane (ppb)	700	1430	1870

Note that carbon dioxide is measured in parts per million (ppm) whereas methane is measured in parts per billion (ppb).

Greenhouse gases Carbon dioxide and methane which trap outgoing radiation and warm the planet up (the natural greenhouse effect)

Enhanced greenhouse effect The natural greenhouse effect working in a more powerful way than it should, because of greenhouse gas emissions from human activity

Global warming The increase in the average temperature of the Earth, especially since 1980, caused by human activity

Now test yourself

What human activity produces the most greenhouse gas emissions?

TESTED

Revision activity

Make a spider diagram showing all the sources of greenhouse emissions from human activity.

Exam tip

Make sure you are clear about the difference between the natural and human causes of climate change.

Evidence for global warming

There is good evidence for the increase in greenhouse gas emissions. Since 1960, the amount of CO_2 and other greenhouse gases in the atmosphere has been directly measured. Figure 1.6 shows how CO_2 levels and global temperatures have risen rapidly in the past 100 years.

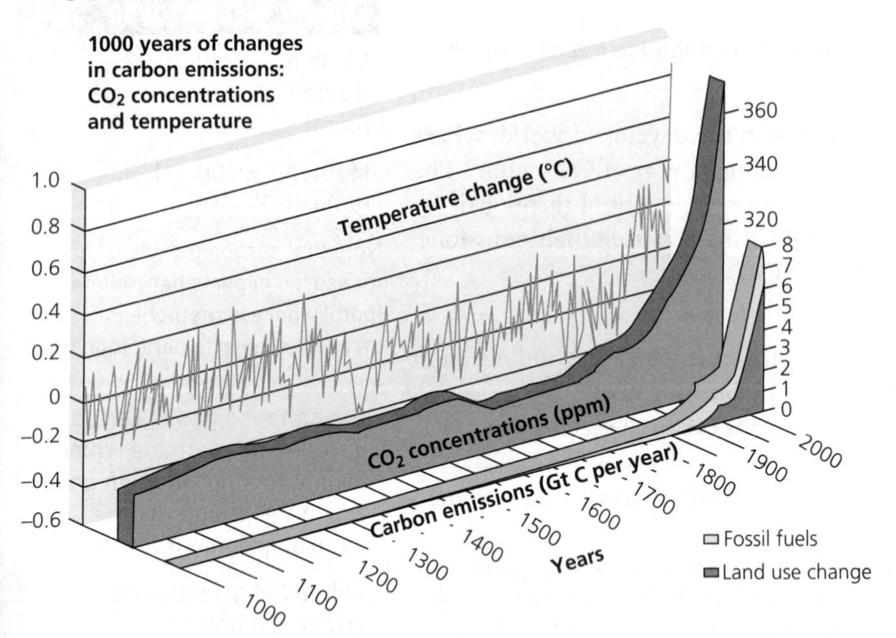

Figure 1.6 How CO_2 levels and global temperatures have risen rapidly in the past 100 years.

There is also growing evidence of change as a result of global warming (see Table 1.3).

Future global warming will have both positive and negative consequences in both developed and developing countries (see Table 1.4).

Table 1.3 Changes as a result of global warming

Sea level rise	Sea level has risen by about 200 mm since 1870 (about the width of this page). This is because: • as temperatures rise, ocean water expands in volume (called **thermal expansion**) • of melting ice sheets and glaciers on land, for example the ice sheets of Greenland and Antarctica and glaciers in Himalayas
Warming oceans	The temperature of ocean surface waters is about 0.5 °C warmer than in 1980, which is enough to begin to affect the distribution of fish species
Global temperature	Global temperature in 2016 was about 0.65 °C higher than in 1980, and about 1 °C higher than in 1920; some places like the Arctic were 2–3 °C warmer
Arctic sea ice	Floating sea ice cover in the Arctic Ocean has shrunk by about 35 per cent since 1979, due to warmer summers
Extreme weather	There is evidence that flooding is becoming more common as storms get more frequent. The UK experienced severe flooding in 2007, 2012, 2014 and 2015

Thermal expansion
An increase in the volume of water as a result of increasing temperature

Extreme weather
Any weather event that poses a risk to human life or property

Table 1.4 Positive and negative consequences of future global warming

Location	Environmental impacts	Economic impacts
The UK (developed country)	New bird and animal species migrate to the UK while others will disappear Changes to fish species in the sea as temperature rises, for example cod moving north out of the North Sea Increased storminess means more erosion on coasts like Holderness	Cost of protecting low-lying areas such as London and East Anglia from rising sea levels Costs to the National Health Service of health problems caused by frequent heatwaves like summer 2003 More frequent and costly flood events such as summer 2007 and winter 2012 Domestic tourism could increase but the Scottish skiing industry is likely to disappear Farmers will need to change crops (from potatoes and wheat to maize and grapes) and irrigation costs could rise
Bangladesh (developing country)	More frequent and/or stronger cyclones in the Bay of Bengal Rising sea levels erode the country's vital coastal mangrove swamps North-west Bangladesh could become more prone to drought	Flooding could become more common with increased rainfall and meltwater from Himalayan glaciers, destroying crops and homes Ten per cent of the land could be lost to rising sea levels, leaving people landless and short of food; ten per cent of people live less than 1 m above sea level Severe water shortages, if the monsoon rains fail, could lead to widespread famine

Now test yourself

TESTED

What is the main cause of the rise in global sea levels?

Revision activity

Rank the consequences of global warming shown in Table 1.4, from most to least serious.

Exam tip

Remember that consequences, impacts and effects can be both positive and negative.

Future projections

A key issue with global warming is the uncertainty about how our climate will change in the future. **Projections** of global sea level rise by the year 2100 range from 20 cm to over 100 cm. Figure 1.7 shows that by 2100 the global average temperature could by anywhere from 1 to 5 °C higher than today.

Projections Estimates of change in the future; the further into the future, the wider the projections become

There are both human and physical reasons for this uncertainty. These relate to the size of future greenhouse gas emissions and how the Earth's climate system might react to higher future emissions (see Table 1.5, page 14).

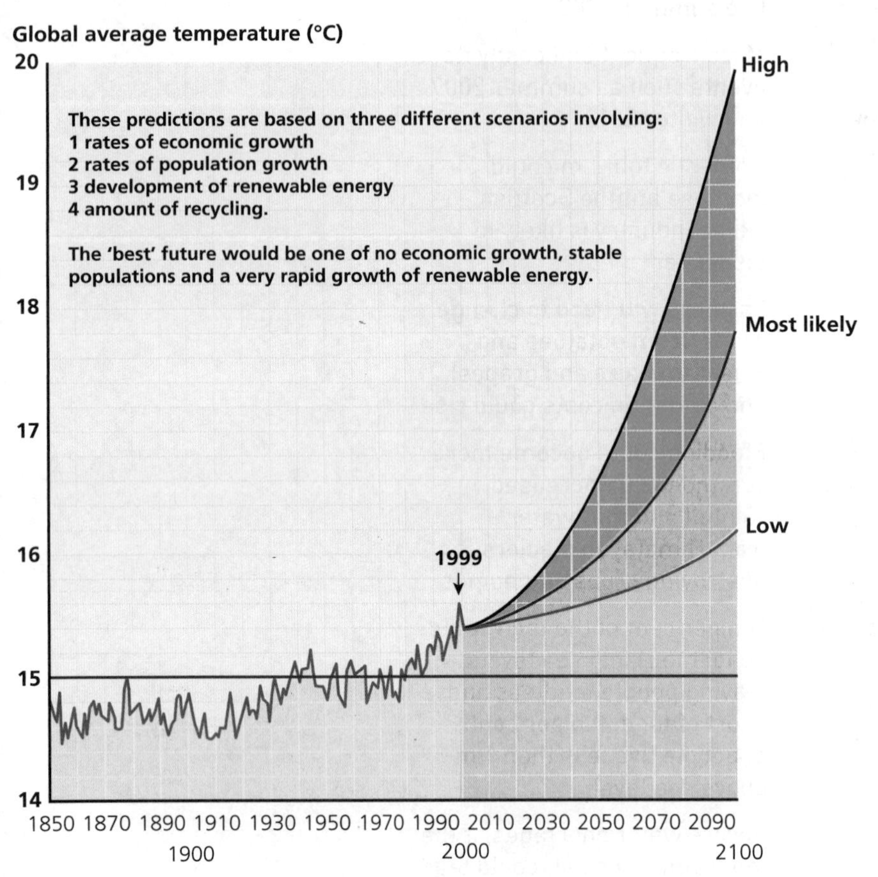

Figure 1.7 By 2100 the global average temperature could be anywhere from 1 to 5 °C higher than today.

Table 1.5 Human and physical reasons for uncertainty

Human reasons for uncertainty	Physical process leading to uncertainty
The world's population in the future is unknown. By 2100 it could be only 9 billion or 13 billion	The oceans may absorb a lot of CO_2, however at some point their ability to do this might stop
Emissions per person are linked to wealth, but the future average wealth of people is uncertain	As sea ice and snow cover in the Arctic shrink, more solar radiation is absorbed rather than reflected, leading to rapid warming
Humans might make a strong effort to reduce emissions, or just carry on polluting	The Greenland and Antarctic ice sheets may start to melt much more quickly in the future, rapidly increasing global sea levels
New technologies could replace fossil fuels, so reducing emissions	A warmer world could be cloudier; clouds could reflect more solar radiation back into space, reducing the warming

Now test yourself

State the high and low range of possible global temperature increases by 2100.

TESTED

Exam practice

1 Describe the future projections of temperature shown in Figure 1.7. [2]
2 Explain two human activities that are major sources of greenhouse gas emissions. [4]
3 Assess the impacts of global warming for both developing/emerging and developed countries. [8]

ONLINE

How are extreme weather events increasingly hazardous for people?

Tropical cyclones

REVISED

Tropical cyclones are a type of **low-pressure** weather system and are a major weather hazard in some parts of the world. They are also known as hurricanes (North Atlantic Ocean and East Pacific) and typhoons (Indian Ocean). Despite the different names, these are all the same weather system.

Low pressure Lower mass of air close to the Earth's surface, caused by air rising or convecting into the atmosphere

Source areas Places where tropical storms form over warm oceans, which in some cases turn into tropical cyclones

Coriolis force A rotational force that affects liquids and gases, caused by the Earth's rotation

Characteristics

Tropical cyclones have these characteristics:

- circular in shape, between 100 and 1500 km wide
- rotating, with spiral bands of thick rain cloud (anti-clockwise in the northern hemisphere, clockwise in the southern)
- a central eye 30–50 km wide, with clear skies and almost no wind
- intense wind speeds around the eye, from 120 to over 200 km/h
- low air pressure in the range of 870–900 millibars.

Tropical cyclones form in very specific locations (see Figure 1.8). The energy for a tropical cyclone comes from the warm ocean. This means they happen in the summer and autumn months in the northern hemisphere, when the Sun has moved north of the equator and heated the ocean. As the Sun moves south of the equator, it heats the ocean there, giving a southern hemisphere cyclone season.

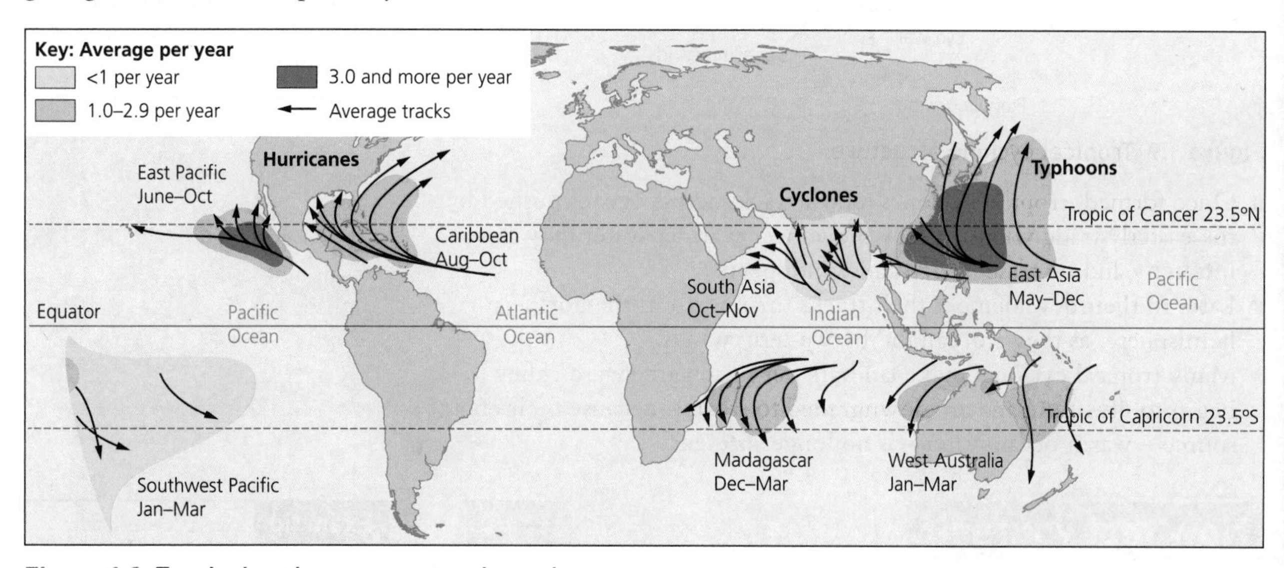

Figure 1.8 Tropical cyclone areas, tracks and seasons.

The **source areas** for tropical cyclones are found 5–15° north and south of the equator, because this is where the ocean is warmest. Source areas are not found:

- further north or south than the tropics, because ocean temperature is not high enough for their formation
- on the equator, or between 0° and 5° north and south, because the force that creates their rotation (the **Coriolis force**) is too weak there.

Revision activity

Using an outline world map, practise labelling the main tropical cyclone locations and their seasons.

Now test yourself

TESTED

In which months are tropical cyclones likely in the northern hemisphere?

Tropical cyclone formation and tracks

Tropical cyclones can form when:

- ocean temperatures are 26.5 °C or higher
- warm air above the ocean rises, leading to strong **convection** and the formation of thick storm clouds.
- As more water is evaporated from the ocean, it rises and **condenses** forming clouds and releasing latent heat into the storm – its energy source.
- Storm clouds combine into one large storm, which begins to rotate.

Figure 1.9 shows the structure of a tropical cyclone. The intense, spiralling upward flow of warm, moist air has to be balanced by a flow of air down towards the ground. This happens in the central eye of the tropical cyclone.

Convection The movement of warm, moist air upwards into the atmosphere (creating low air pressure)

Condensation Water vapour (gas) turning to liquid droplets as air rises and cools, forming clouds

Intensify The process of tropical cyclones growing in size, wind speed and rainfall levels as they gain energy

Dissipation The final stage of a tropical cyclone when, having no energy source, wind speeds, rotation and rainfall all reduce

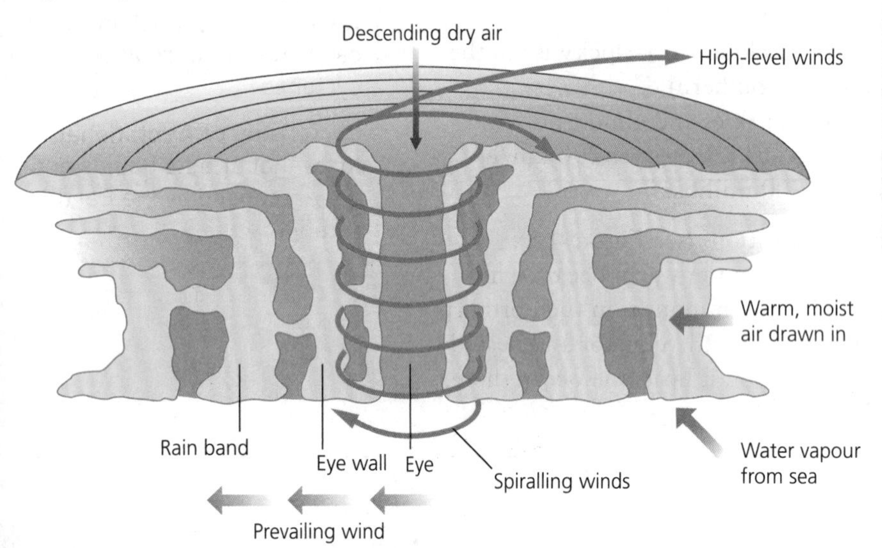

Figure 1.9 Tropical cyclone structure.

Exam tip

Remember that tropical cyclones only form when a number of conditions are met, not just high ocean temperatures.

- Once formed, tropical cyclones follow tracks across oceans pushed by the easterly trade winds. If they move over warmer water they can **intensify** and increase in size and wind speed.
- Later in their development their tracks turn north in the northern hemisphere, as they are 'caught' by westerly winds.
- Many tropical cyclones make landfall; very soon afterwards, they begin to **dissipate** and are downgraded to storms, because their energy source – warm ocean water – is no longer present.

Now test yourself

TESTED

What ocean temperature is needed for a tropical cyclone to form?

Exam tip

You need to be able to explain how tropical cyclones form in a logical sequence from start to dissipation.

Exam practice

1 Describe the distribution of tropical cyclones shown in Figure 1.9. [2]
2 Explain how tropical cyclones form. [4]

ONLINE

Tropical cyclone hazards

REVISED

Tropical cyclone intensity is measured using the Saffir–Simpson scale (see Table 1.6).

Table 1.6 The Saffir–Simpson scale

Category	Sustained winds (km/h)	Storm surge height (m)	Damage
1	119–153	1.0–1.5	Minimal
2	154–177	1.6–2.4	Moderate
3 (major)	178–208	2.5–3.6	Extensive
4 (major)	209–251	3.7–5.4	Extreme
5 (major)	252+	5.5+	Catastrophic

Tropical cyclones making landfall at Category 3 or higher have the potential to cause enormous damage. There are a number of different physical hazards that happen when cyclones make landfall:

- Intense winds – these can uproot trees, blow the roofs off buildings, and bring down power and telephone cables; and flying debris can be a fatal hazard to people.
- Intense rain – this can lead to flooding, washing away bridges and roads, and causing soil erosion in fields.
- **Landslides** – intense rainfall can weaken steep slopes, leading to landslides burying houses and blocking roads.
- Powerful waves – waves driven onshore by strong winds can cause coastal erosion and smash sea defences and flood defences.
- **Storm surge** – the low air pressure inside a tropical cyclone allows the sea surface to rise by 10 mm for every 1 millibar drop in air pressure; when the cyclone makes landfall it can cause widespread coastal flooding as the sea floods coastal land.

Most of the damage caused by tropical cyclones is caused by flooding as a result of the storm surge. If people have not been evacuated, they risk being drowned as well as being hit by flying debris, collapsing buildings and raging flood waters.

Tropical cyclones can have widespread impacts (see Table 1.7).

Landslides Occur on slopes when rock or soil collapses downwards, destroying property and blocking roads

Storm surge A temporary rise in sea level at the coast, which leads to widespread flooding

Revision activity

Look at Table 1.7 on tropical cyclone impacts: which are more likely or less likely in developed versus developing countries?

Table 1.7 The widespread impacts of tropical cyclones

Economic impacts	Social impacts	Environmental impacts
Loss of crops, caused by widespread flooding Businesses destroyed, and forced to close during evacuations Repairs to roads, bridges and flood defences after the cyclone	Deaths and injuries Psychological trauma In developing countries, homelessness when people lose everything Families split up during evacuation	Widespread damage to coral reefs, mangrove swamps and other coastal ecosystems Farmland polluted by saltwater from storm surges Increased coastal erosion

Now test yourself

TESTED

What tropical cyclone hazard causes coastal areas near the sea to flood?

Exam tip

Remember that tropical cyclones cause multiple hazards, not just very strong winds.

Cyclone vulnerability

Not all countries are equally vulnerable to the impacts of tropical cyclones:

- Low-lying coastal areas, often with flat fertile farmland or coastal cities such as New Orleans in the USA and Dhaka in Bangladesh, are the most likely to be flooded.
- Coastlines often contain industry and tourism infrastructure which means large economic losses if hit.
- In places where coral reefs have been damaged and mangrove swamps cut down, the natural protection these ecosystems provide has been lost.
- Some locations are hit again and again, such as Haiti and the Philippines, because they are on cyclone tracks and can expect several tropical cyclones each year.

In developing countries like Bangladesh and the Philippines, people are especially vulnerable due to poverty. This is most true of rural subsistence farming families who grow their own food to eat. They have no insurance or financial safety net, and live in high-risk places in flimsy buildings, and in isolated locations which are often the last to receive aid – and the warning that a cyclone is coming.

Now test yourself

TESTED

Why are cities like New Orleans and Dhaka very vulnerable to coastal flooding?

Exam practice

1 Explain how tropical cyclones produce multiple hazards. [4]

2 Explain two reasons why some areas are more vulnerable to tropical cyclones than other areas. [4]

ONLINE

Cyclone preparation and response

REVISED

Much can be done to prepare for and respond to tropical cyclones, but the effectiveness of preparation depends partly on how developed a country is. Developing countries are often more vulnerable.

Table 1.8 Preparation and responses to tropical cyclones in a developed and a developing country

USA (developed country)	Bangladesh (developing country)
Satellite technology, airborne cyclone monitoring flights and weather radar technology are all used to predict the landfall location and intensity of tropical cyclones **Weather forecasting** is increasingly accurate Warnings are broadcast on TV and radio, and **evacuation** takes place along signposted routes away from the coast to inland shelters – open schools and sports centres Costly storm-surge defences, called levees, can withstand the flooding of all but the worst storms When places are badly hit, FEMA (the Federal Emergency Management Agency) mobilises resources to repair damage and temporarily rehouse people If a large disaster has occurred, money from national government is released to local government to help with the repair bill Many people have their own home and business insurance, meaning they can recover	Warnings are issued on radio, on TV and increasingly via text message However, people cannot be evacuated far, if at all, due to lack of transport and roads Many people move to a cyclone shelter; there are around 500 of these concrete shelters built on stilts These shelters cost £250,000 each – a lot of money for a developing country Cattle are sometimes put on fenced embankments called kilas in the hope they will survive the storm There are not enough shelters for everyone and shelters do not protect people's homes or crops After a major cyclone, help from the government is often boosted by foreign aid: overseas governments and charities donate money, food and technical equipment to help people to rebuild their lives Longer term, the development of rice varieties that can withstand flooding by saltwater would help farmers hugely as their crops would survive

Revision activity

Make up some flash cards for your own named examples of tropical cyclones, using 'impacts' and 'responses' to organise your knowledge.

Exam tip

Longer questions in the exam will expect you to be able to discuss tropical cyclone impacts in both developed and developing countries.

Now test yourself

TESTED

What is FEMA responsible for in the USA?

Exam practice

Assess the reasons why preparation for tropical cyclones is better in some countries than in others. [8]

ONLINE

Satellite technology Satellites orbit the Earth and provide images from above tropical cyclones that are used to predict their track

Weather forecasting Computers, weather data and satellites are used to predict the weather up to ten days ahead

Evacuation The temporary movement of people to a safe location, to reduce their vulnerability

Why do the causes, impacts and management of tectonic activity vary with location?

Earth's layered structure

REVISED

The Earth is made up of several different layers. The diameter of the Earth is about 13,000 km and the outer layer – the crust – is between 6 and 80 km thick. The upper part of the mantle and crust are known as the **lithosphere**. Both temperature and density increase with depth. There are two types of crust, which make up Earth's mobile **tectonic plates**:

- Continental crust (low density) makes up most of the land area of the Earth. It is dominated by rocks that cooled below the surface, such as granite. It is between 25 and 80 km thick.
- Oceanic crust (high density) is much thinner – between 6 and 8 km thick – and made up of rocks such as basalt.

The density difference between the two types of crust is important in plate tectonics.

Lithosphere The rigid outer layer of the Earth; its upper part is called the crust

Tectonic plates Slabs of lithosphere that fit together like a jigsaw across the Earth's surface and constantly move

Asthenosphere A partly molten layer beneath the lithosphere on which the lithosphere slides

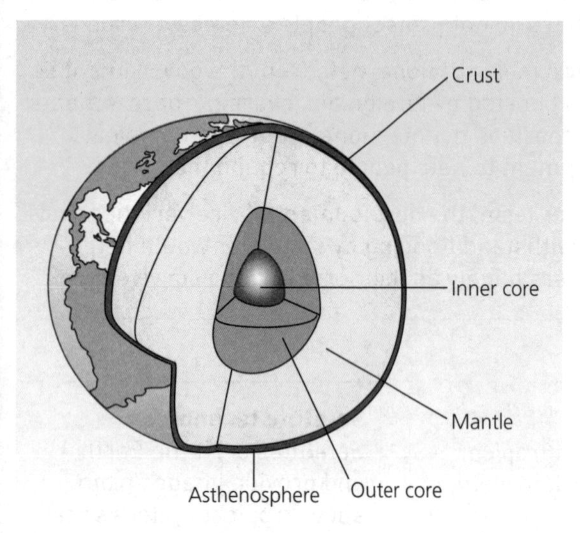

Figure 1.10 The Earth's layered structure

Revision activity

Sketch a rough cross-section diagram of the Earth's interior and label the different layers.

Exam tip

Good exam answers will contain data on the Earth's layers, not just vague phrases like 'not very dense'.

Table 1.9 The layers of the Earth.

Layer		Physical state	Composition	Temperature (°C)	Density (g/cm³)
Lithosphere	Continental crust	Solid	Granite	Up to 900	2.7
	Oceanic crust	Solid	Basalt	Up to 900	3.0
Mantle	**Asthenosphere**	Partially molten	Peridotites	1000–1600	3.4
	Mantle	Solid	Silica-based minerals	1600–4000	4.5–5.5
Core	Outer core	Liquid	Iron/nickel	4000–5000	9.9–12.2
	Inner core	Solid	Iron/nickel	4000–5000	12.6–13.0

Now test yourself

TESTED

Which of Earth's layers is liquid?

Plate tectonics

Tectonic plates move very slowly, at a rate of about 2 cm per year. The cause of this movement is heat generated in the Earth's core. High temperatures in the core are caused by radioactive decay of the elements uranium and thorium. This heat creates rising limbs of material in the mantle, called **convection currents**. These cool and spread out as they rise before sinking again – just like a lava lamp. Some of this rising and falling material moves in sheets, creating movements in the crust above it, which is pulled apart to form new crust. In other places it rises as columns, creating hotspots. Plate motion is caused by three forces:

- Plates are pushed along by convection currents moving sideways just below the crust.
- Old, dense oceanic plate sinks at convergent margins, pulling the plate down into the mantle; this is called slab-pull.
- The elevated parts of divergent boundaries create a force called gravitational sliding.

Convection currents
Circular movements of heat within the Earth's core and mantle, which drive plate-tectonic motion

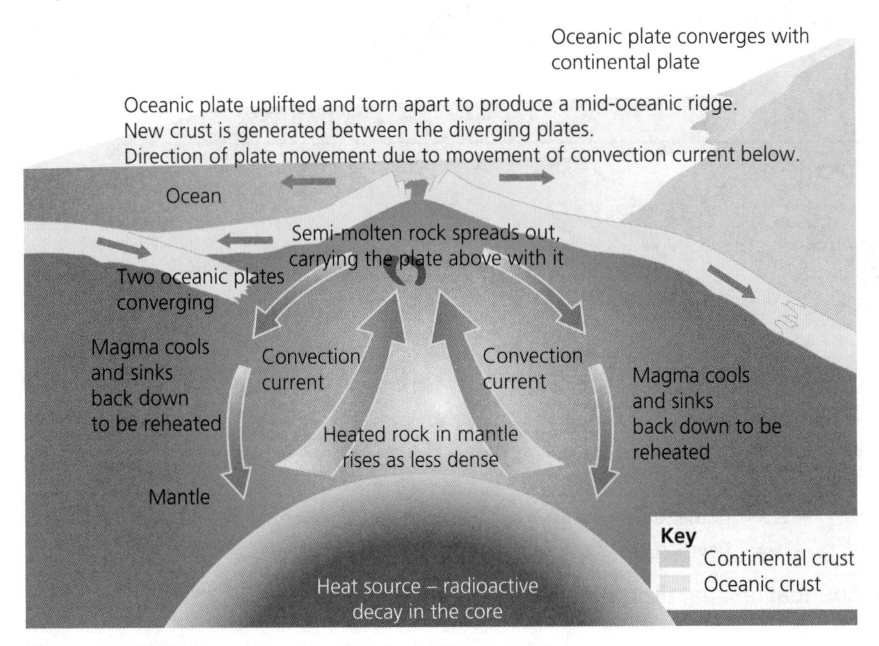

Figure 1.11 Convection currents in the mantle.

Exam practice

1 Describe the structure of the Earth's interior shown in Figure 1.10. [2]
2 Explain the causes of tectonic plate movement. [4]

ONLINE

Plate boundaries

REVISED

Plate boundaries are the edges of plates where one plate meets another. Most earthquakes and volcanoes occur here. There are three types of plate boundary:

- conservative boundaries, where two plates slide past each other
- convergent (or destructive) boundaries, where two plates collide
- divergent (or constructive) boundaries, where two plates move apart from each other.

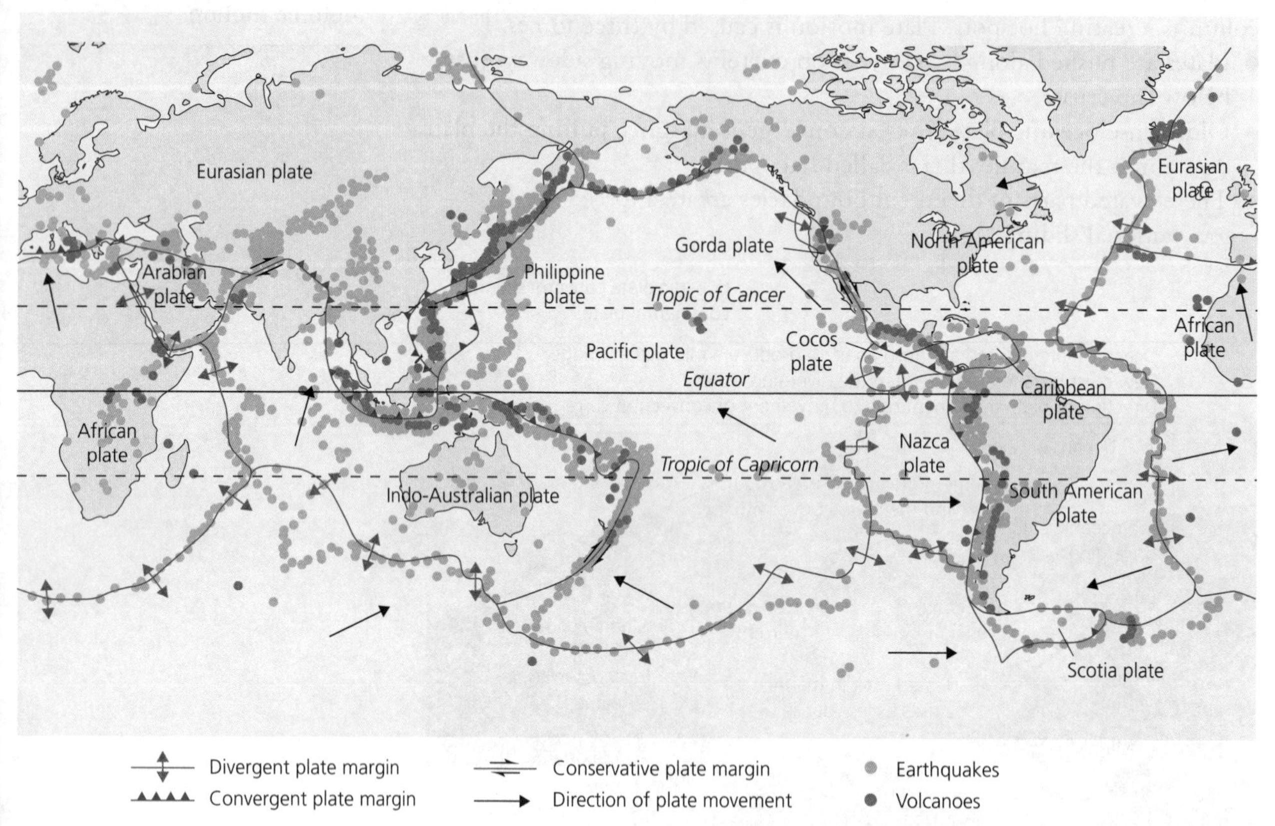

Figure 1.12 **The main plates, plate boundaries, earthquake zones and active volcanoes.**

Divergent margins

- These are formed by rising magma splitting up continental crust and forming new oceans, such as the Atlantic.
- This is happening in east Africa today in its continental rift zone.
- In the middle of oceans, divergent margins create new oceanic crust and sub-marine mountain chains.
- The mid-Atlantic ridge is a divergent margin sometimes visible above sea level, as in Iceland.

Convergent margins

- In some places, such as where the Nazca plate meets the South American plate, oceanic plates collide with continental plates.
- When this occurs, the denser basaltic oceanic plate sinks beneath the continental plate.
- This process is known as **subduction** and creates a very deep ocean trench near the line of contact between the oceanic and continental plates.
- As an oceanic plate is subducted into the mantle it is subjected to increased pressure and temperature.

Subduction The process whereby a dense oceanic plate sinks back into the upper mantle

- These conditions cause some low-density minerals in the plate to melt, forming magma, which rises to the surface to form volcanoes.
- As a result, long chains of volcanoes, known as volcanic arcs, are located above subducted plates, usually above the location where the plate has reached a depth of about 100 km.
- In rare cases, two continental plates collide creating **fold mountains**, for example the Himalayas; this is because the low-density continental plates cannot be subducted into the mantle.

Fold mountains Chains of high mountains produced when tectonic plates collide and the continental crust crumples and is forced upwards

Faults Major cracks in the crust, and often the place where earthquakes start when there is movement on the fault

Conservative margins

Where plates slide past each other or move in the same direction but at different speeds:

- no crust is formed or destroyed, and volcanoes do not form
- great strain builds up along the boundary and cracks in the crust, called **faults**, generate powerful earthquakes when they move
- a famous example of this boundary type is the system of faults along the west coast of the USA, the best known of which is the San Andreas fault.

Hotspots

- These are places where there is a rising plume of heat from the mantle in the form of an isolated column.
- The heat generates magma in the upper mantle, which erupts as basaltic volcanoes.
- The tectonic plate moves slowly over the stationary heat plume, creating a chain of volcanic islands which begin as active volcanoes, but become extinct when they move away from the plume.

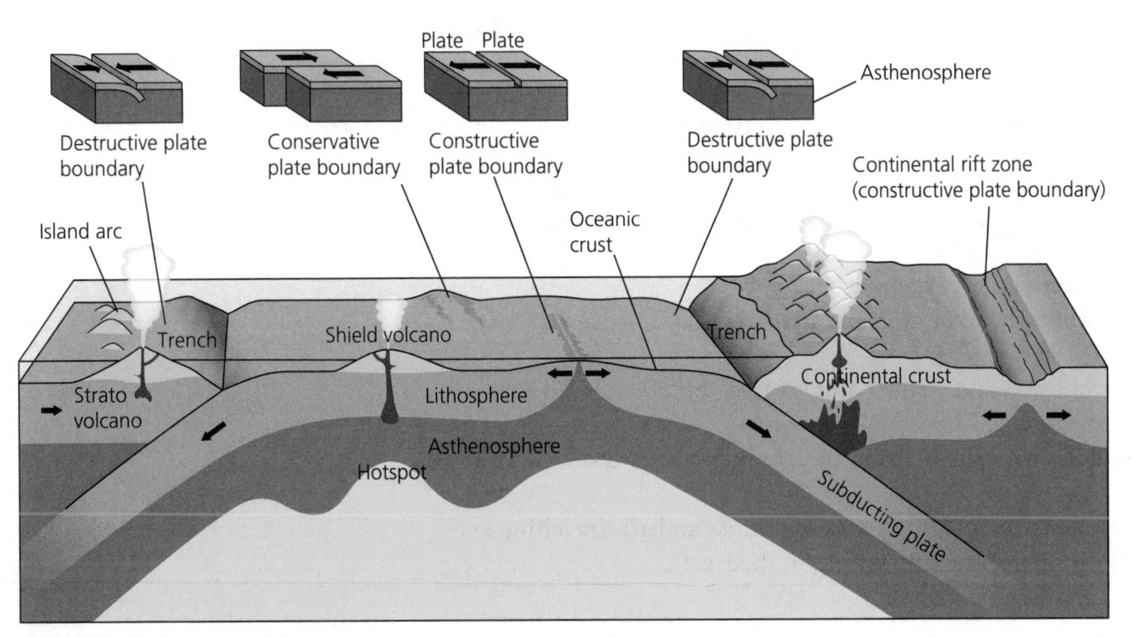

Figure 1.13 Plate boundaries.

Now test yourself

TESTED

Which type of plate margin does subduction happen at?

Exam tip

Make sure you can recognise each type of plate boundary on a cross-section diagram, as these are often used in exams.

Volcanic and earthquake hazards

There are two basic types of tectonic hazard:

- Volcanic eruptions happen when molten **magma** beneath the Earth's crust pushes up through fissures and vents, and erupts at the surface as **lava**, ash and gas.
- Earthquakes are caused when movement happens along faults (cracks) within tectonic plates, because of a build-up of stress over time.

Most eruptions and earthquakes happen at plate margins.

Magma Underground molten rock

Lava Molten rock erupted at the surface

Explosivity A measure of how violent an eruption is, how much material is erupted and how high the eruption cloud is

Volcanoes

Volcanic eruptions come in two types: explosive and effusive. How dangerous a volcanic eruption is can be measured using the volcanic **explosivity** index or VEI scale. Composite volcanoes found on convergent plate boundaries are by far the most dangerous type.

Table 1.10 Different types of volcanic eruptions

Volcano type	Shape and size	Magma/lava type	Explosivity	Examples
Composite cone or strato (convergent boundary)	Steep-sided cone shape Small area Alternating layers of ash and lava	Andesitic Viscous ('sticky'), low-temperature magma with high gas and silica content that does not flow easily	VEI 4–8 Infrequent eruptions Pressure builds up over decades or centuries, eventually erupting explosively	Mt Pinatubo (Philippines), Mt Merapi (Indonesia) and Mt Sakurajima (Japan)
Shield (divergent boundary or hotspot)	Gently sloping Very large area Made of solidified lava	Basaltic Non-viscous ('runny'), high-temperature magma with low gas and silica content that flows rapidly and easily	VEI 0–3 Frequent, sometimes almost continuous, eruptions of very hot lava	Mauna Loa (Hawaii, USA) Mt Nyiragongo (DR Congo)

Composite volcanoes are dangerous because:

- when they erupt, the explosion and its effects can impact on a large area
- as well as lava flows, ash, tephra and volcanic bombs can fall across a wide area, destroying buildings and crops
- volcanic mudflows, called lahars, are a frequent hazard on the volcano's slopes
- in the worst cases, pyroclastic flows of hot gas and ash travelling at 200 km/h can incinerate everything in their path.

Earthquakes

Earthquakes are a huge instant release of stored energy. Earthquake destruction depends on several factors:

- the **magnitude** of the earthquake, measured using the Richter scale
- this scale measures the energy released; a magnitude greater than 6.0 is likely to be serious, and above 8.0 devastating in a populated area
- the depth in the crust at which the earthquake starts (its focus or hypocentre)
- shallow earthquakes at less than 30 km depth are the most destructive because earthquake waves reach the surface quickly with little loss of energy.

Magnitude The size of an earthquake, measuring the amount of energy released

Tsunami Destructive waves in the ocean, usually caused by earthquakes under the seabed

Large volcanic eruptions and earthquakes can both cause a **tsunami** as a secondary hazard. Tsunamis are a series of very destructive ocean waves. Sub-sea earthquakes can displace the seabed and this causes a tsunami. The waves travel across oceans at speeds up to 900 km/h. When they strike land, waves can be 20–30 m high and flood far inland. The Indian Ocean tsunami on 26 December 2004 killed over 250,000 people in fourteen countries.

Now test yourself

TESTED

Which type of volcano is more dangerous to people: composite or shield?

Revision activity

Use Table 1.10 on shield and composite volcanoes as a guide to make notes on your own case studies.

Exam practice

1 Define the term 'hotspot'. [2]
2 Explain the process of subduction at a convergent plate boundary. [4]
3 Explain two reasons why some volcanoes have high explosivity. [4]

ONLINE

The effects of tectonic hazards

REVISED

A number of factors control how severe the impacts of earthquakes and volcanic eruptions are. These include the magnitude of the event, the time of day, the level of preparedness and the quality of the emergency services. Usually, the less developed the country the greater the impact on people. On the other hand, because the population in developed countries insure their property and businesses, the economic cost of the disaster is often higher.

Primary and secondary impacts

Impacts are often divided into:

- Primary impacts – the immediate effect of an earthquake or eruption on property and people.
- Secondary impacts – the impact on property and people of a hazard after it has finished. Lack of shelter and basic supplies, as well as fires, are frequent secondary effects.

These impacts can be contrasted for developed and developing world examples (see Table 1.11, page 26).

Table 1.11 Earthquake impacts in a developing and a developed country

Earthquake example and details	Earthquake details	Impacts
Port-au-Prince, Haiti, January 2010 (developing country)	Magnitude 7.0 Depth 13 km Struck at 5p.m.	Primary: 316,000 deaths and 300,000 injured. Total economic losses of £8.5 billion. Poverty and slum housing made people very vulnerable to building collapse and secondary impacts such as cholera Secondary: over 7000 people killed. An estimated 1 million people were made homeless; damage to roads and ports stopped trade; cholera spread due to lack of clean water and sanitation
Canterbury, New Zealand, September 2010 (developed country)	Magnitude 7.1 Depth 10 km Struck at 4.30a.m.	Primary: no deaths, about 100 injuries. Total damage to property about £1.8 billion. Deaths were low because most people were asleep and buildings were strong structures Secondary: a major aftershock, magnitude 6.3, occurred in February 2011 in nearby Christchurch, killing 185 people

Volcanic eruptions have smaller impacts than earthquakes, but they can still be significant. Mt Merapi in Indonesia (an emerging country) is a composite volcano that erupted in 2006 and again in 2010. Merapi is a dangerous volcano with a large population living nearby (see Table 1.12).

Table 1.12 Impacts of the 2010 Mt Merapi eruption

Impacts	Social	Economic
Primary impacts	360,000 people evacuated from the area; some refused to go and others returned during the eruption 275 people were killed, mostly by scalding hot ash and gas in pyroclastic flows 570 people were injured	Several villages were destroyed, and damage to crops from ash fall was widespread About 2000 farm animals were killed Many aircraft flights in the area were cancelled due to the ash cloud
Secondary impacts	An area 10 km around the volcano was declared a danger zone and 2600 people were not able to return to it Thousands spent weeks living in cramped emergency centres	1300 hectares of farmland were abandoned Economic losses of $600 million due to severely reduced farming and tourism income

Now test yourself

TESTED

How many people were made homeless by the 2010 Haiti earthquake?

Exam tip

It is very easy to confuse primary and secondary impacts in the exam, so make sure you learn both carefully and know the difference.

Managing volcanoes and earthquakes

It is possible to reduce the impacts of earthquakes and volcanic eruptions by careful **management**. However, this is generally:

- easier for volcanic eruptions, as more warning can be given
- more widespread in the developed world, and less so in emerging and developing countries that have fewer financial and technical resources.

Exam tip

In the exam, read the question carefully. Students often answer a question about volcanoes using an earthquake example by mistake!

Table 1.13 Managing earthquake and volcanic hazards

Technique	Managing earthquake hazards	Managing volcanic hazards
Prediction	Earthquakes cannot be predicted, despite areas of high risk being well known	Most eruptions are predicted, thanks to monitoring equipment which measures gas emissions, ground temperature and the movement of magma below ground
Preparation	In developed countries like the USA, land-use zoning prevents building in very high-risk areas Costly earthquake-proof (aseismic) buildings are common in the USA and Japan but not for ordinary people's homes In Japan, earthquake drills teach people how to act during an earthquake and some people keep an 'Earthquake Kit' of emergency supplies at home	In the USA, evacuation routes are signposted and sirens sound if eruptions are imminent However, evacuating big cities like Naples in Italy could be problematic In rural, isolated areas of developing countries not everyone might get a warning Most buildings cannot be made 'volcano proof'
Long-term planning	In the USA and Japan, fire, police and search and rescue services are highly trained and well funded Developing countries often have to rely on rescue teams arriving from the developed world	Educating people about volcanic risk is important and often carried out by non-governmental organisations (NGOs) in developing countries
Short-term relief	Aid from NGOs and organisations such as the United Nations is often very important for shelter and food supplies in the months after a major earthquake in a developing country	Providing relief and aid is often easier than for an earthquake because people have already evacuated Deaths and injuries are lower than for a major earthquake

Now test yourself

TESTED

Which tectonic hazard can be predicted: earthquakes or volcanic eruptions?

Exam practice

1 Suggest two reasons why the impacts of earthquakes are often worse in the developing world. [4]

2 Assess the extent to which the impacts of earthquakes can be reduced by preparation and relief. [8]

ONLINE

Management Ways in which the risk to people can be reduced and property and/or lives saved

Prediction The ability to say both when and where a hazard will strike, with enough accuracy to warn and evacuate people

Preparation Any management used before the hazard strikes, designed to make people less vulnerable and therefore save lives and property

Topic 2 Development dynamics

The scale of global inequality and how to reduce it

Defining and measuring development

REVISED

Development means making progress so that people's lives improve. Lives can improve in different ways, for instance, having more money, living longer or knowing your children can get an education. Development is defined in different ways (see Table 2.1).

Table 2.1 Definitions of development

Definition of development	Explanation
Economic development	Increasing numbers of people working in secondary and tertiary employment sectors – leading to rising per capita incomes
Social development	Rising life expectancy, better healthcare and access to education; improved equality for women and minorities – leading to improved quality of life
Political development	Improving **political freedom**, the right to vote, a free press and freedom of speech, reduced **corruption** – leading to greater control over who governs you

Contrasting measures

Development often focuses on economics. If people are getting richer, their lives must be improving, but this is not always the case:

- Living in polluted, congested cities might worsen health and increase stress.
- Incomes could rise, but so could the cost of living (food, water, housing), especially in cities.

Broadly speaking, there are three types of countries, with each at a different stage of development. These are developing, emerging and developed countries (Figure 2.1).

Political freedom The right to vote and express ideas freely, and the freedom of the press to report stories

Corruption Illegal activity, such as bribery and avoiding taxes, or stealing government funds

Exam tip

Make sure you know the pattern of developing, emerging and developed countries shown in Figure 2.1.

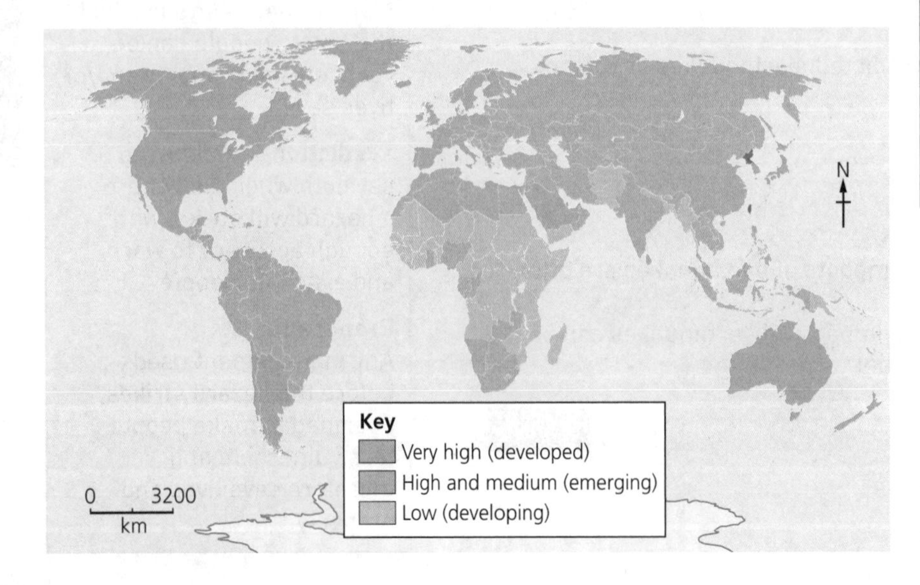

Figure 2.1 Developing, emerging and developed countries based on their human development index.

Development progress can be measured in different ways, each way having advantages and disadvantages (see Table 2.2).

Table 2.2 Measures of development progress

Development measure	Type of development	Advantages	Disadvantages
Gross domestic product (GDP) per capita	Economic development	A useful 'headline' figure Easy to understand	Ignores regional differences Does not measure inequality
Human development index (HDI)	Social development (combines income, life expectancy and education)	An index is more accurate than a single measure Tracks progress over time	Data is hard to collect and calculate Ignores issues like freedom
Gini coefficient	**Inequality** (the difference between richest and poorest)	Indicates how fair a country is	Does not indicate how wealthy or poor a place is
Corruption perceptions index	Measures dishonesty and the use of bribery	High levels of corruption can hold a country back	The data is complex and may not be accurate

Now test yourself

TESTED

Which three measures are combined in the human development index?

Exam tip

You need to be able to assess the advantages and disadvantages of different ways of measuring development for the exam.

As can be seen in Table 2.3, there are very large differences between countries using these different measures of development.

Table 2.3 Measures of development progress: contrasting countries

Country	Nominal GDP per capita (US$)	HDI (1 = best, 0 = worst)	Gini coefficient (0 = most equal, 100 = most unequal)	Corruption perceptions index (100 = best, 0 = worst)
Haiti (developing)	800	0.29	61	20
Mexico (emerging)	11,200	0.58	48	30
Sweden (developed)	48,900	0.85	27	88

Revision activity

Learn some development level data for a developing, an emerging and a developed country so you can quote it in longer answers.

Gross domestic product (GDP) The value of all of the goods and services produced within a country in a year

Inequality How unfair a society is, in terms of the income and opportunity gap between rich and poor, or between different genders or ethnic groups

Development and population

A major difference between countries at different levels of development is **demographics**. Development level affects:

- birth, death and **infant mortality** rates – which all change as a country develops
- the structure of a country's population pyramid – as the number of people in each age group changes, so does the shape of the pyramid.

Demographics Features of a country's population such as different age groups and the number of men and women

Infant mortality The number of babies that die before one year of age (sometimes five years) per 1000 born

Table 2.4 Demographic data in contrasting countries

Population pyramid	Demographic data	Explanation
Haiti (developing) 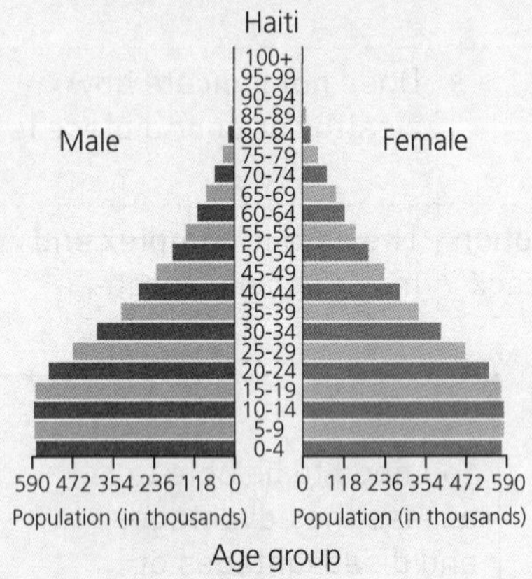	Birth rate: 23 per 1000 people Death rate: 8 per 1000 people Fertility rate: 3.2 children per woman **Maternal mortality**: 350 deaths per 100,000 births Infant mortality: 48 per 1000 live births	Rapid population growth Lack of access to contraception, and a need to have children to work keep fertility rate and birth rate high Poverty, poor healthcare and lack of sanitation mean infant and maternal mortality rates are high. Most people are aged under twenty
Mexico (emerging) 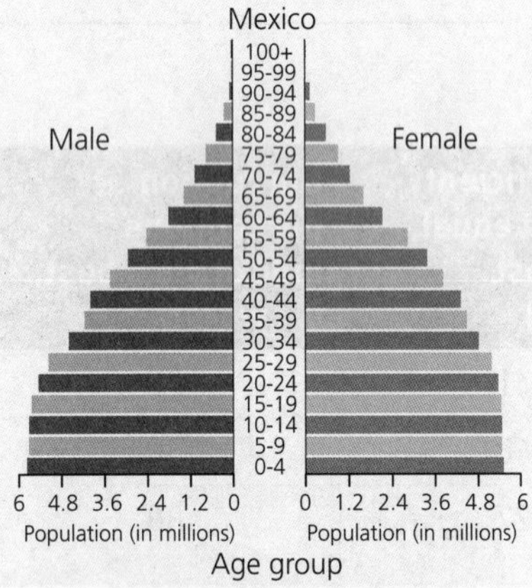	Birth rate: 19 per 1000 people Death rate: 5 per 1000 people Fertility rate: 2.2 children per woman Maternal mortality: 50 deaths per 100,000 births Infant mortality: 12 per 1000 live births	Slowing population growth Improving healthcare, education and incomes mean children are needed less as economic assets so fertility rates fall and life expectancy increases. The working age and older population are growing
Sweden (developed) 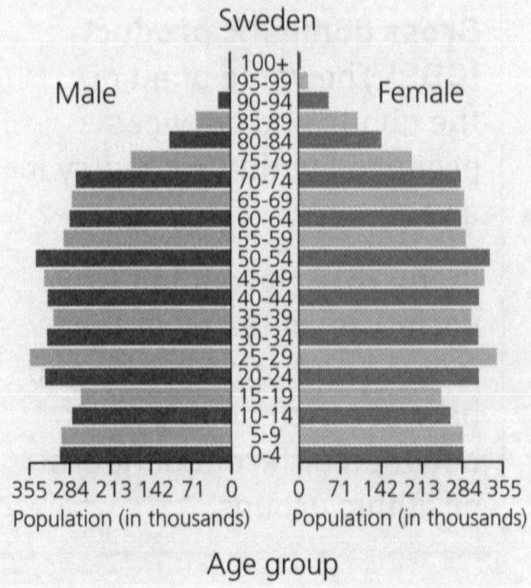	Birth rate: 12 per 1000 people Death rate: 9 per 1000 people Fertility rate: 1.8 children per woman Maternal mortality: 4 deaths per 100,000 births Infant mortality: 3 per 1000 live births	Stable, or even declining, population Birth rate and fertility rate are low as many women work and small families are normal. Excellent healthcare reduces infant and maternal mortality to very low levels But the population is ageing, with few young people

Maternal mortality The number of women who die giving birth per 100,000 live births

Revision activity

Practise sketching population pyramids and labelling their key features.

Exam tip

Sometimes sketching and labelling a population pyramid in the exam is faster than trying to describe it.

Now test yourself

TESTED

Which type of country has the largest proportion of people aged 0–15 in its population pyramid?

Exam practice

1 Using Figure 2.1, describe the distribution of developing countries. [2]
2 Explain two ways that development level can be measured. [4]
3 Describe the shape of Haiti's population pyramid. [2]
4 Explain why fertility rates fall as a country develops. [4]

ONLINE

Global inequalities

REVISED

The world is very unequal in terms of development, as Figure 2.2 shows. On a global scale, the richest twenty per cent of people own 85 per cent of the world's wealth, whereas the poorest twenty per cent of people own less than two per cent of the wealth. A 2016 report by Oxfam stated that the world's eight richest billionaires had as much wealth as the poorest half of the world – 3.6 billion people.

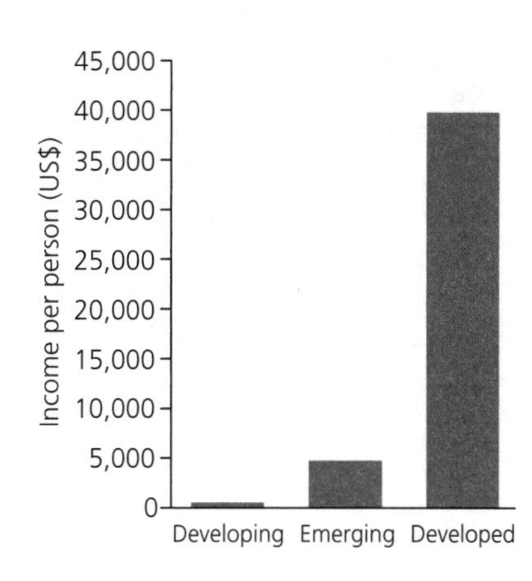

Figure 2.2 Average per capita incomes in developing, emerging and developed countries in 2015.

Causes

There are many reasons for this inequality (see Table 2.5, page 32).

Development can be prevented because some countries have poor **international relations** with other countries. An example is North Korea. Most countries refuse to trade with North Korea because of its brutal dictatorship and its plans to develop nuclear weapons. Without trade, it is hard for North Korea to become wealthier.

International relations The channels of communication, disagreements, treaties and alliances between countries

Table 2.5 Reasons for global inequality

Social reasons	Historical reasons
In developing countries, education is basic and often stops after primary school Girls may not be educated at all Lack of education limits people's future life chances and opportunities to earn a good income Poor healthcare in developing countries means disease affects people's ability to work In many African countries, like Botswana and Uganda, HIV/AIDS is a huge problem and it causes poverty	Many African and Asian countries were once **colonies** of the UK, France and other European countries Colonies supplied their owners with raw materials like coffee, rubber and tin but the colonial powers did little to develop their colonies in terms of infrastructure, education and health In more recent times, newly independent countries that were colonies have continued to supply cheap raw materials to developed countries – earning very little money. This situation is called **neo-colonialism** Some countries are land-locked, with no access to the sea. This makes trade, and therefore development, difficult and expensive
Environmental reasons	**Political reasons**
Some places have an extremely harsh climate, such as the deserts of the Sahara and Middle East Water supply is limited, meaning farming is hard and this limits development The dense tropical forests of Amazonia and Indonesia are difficult to 'open up' to development Isolated, mountain countries like Nepal, Bolivia and Bhutan are hard to develop because of their difficult **topography**	Many developing countries are not democracies They have undemocratic, often corrupt **systems of governance** This means people cannot get rid of a government that is not helping the country to develop Often profits from companies, or even aid from other countries, are stolen and just used to make the people in power richer This corruption means money is diverted from development spending on roads, schools and hospitals

Colonies Areas that were owned and run by an invading colonial power, and were taken by force

Neo-colonialism The idea that developed countries control developing ones indirectly

Systems of governance The way a country is run, for example as an elected democracy or an unelected dictatorship

Topography The shape of the land, whether areas are mountainous or flat

Industrial Revolution A period of rapid development in manufacturing industry, linked to the rise of emerging economies

Revision activity

You may have looked at a case study of an African developing country and why it has struggled to develop. Link your case study to the social, historical, environmental and economic and political factors in Table 2.5.

Development theories

Two theories can help us to understand why countries develop over time. One is W.W. Rostow's modernisation theory (Figure 2.3). Rostow believed that development needed certain 'preconditions' before it could happen. These preconditions included:

- a move from farming to manufacturing industry
- trade with neighbouring countries
- the development of infrastructure such as roads, ports and railways.

Once these happened a country would 'take off' – it would develop secondary industry, begin to export and slowly increase trade and incomes. Rostow's theory is based on what happened in Britain during the **Industrial Revolution**.

Rostow's theory has been very influential. Many countries have tried to get to stage 3 by investing in infrastructure, although they often borrow money to do so, which can lead to unsustainable debt.

Exam tip

Don't confuse historical colonialism with modern-day neo-colonialism.

Now test yourself

Explain how disease could affect people's ability to earn a living.

TESTED

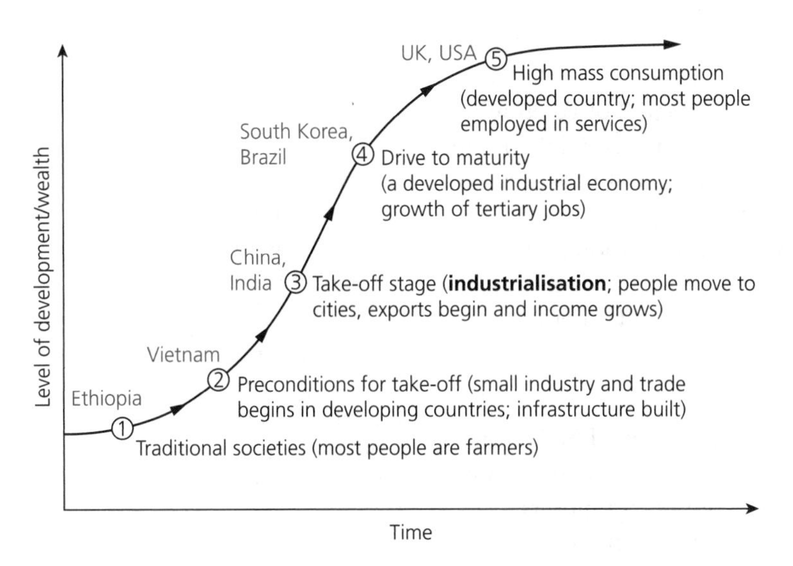

Figure 2.3 The five stages of Rostow's modernisation theory with country examples.

An important question is, why haven't all countries managed to move out of stage 1 or 2? The answer might be provided by **dependency** theory, an idea put forward by the economic historian A.G. Frank. Frank argued that developed countries exploit developing countries and this keeps them in a state of 'underdevelopment' (Figure 2.4).

Dependency Relying on someone else, as when developing countries depend on developed ones for trade

Developed countries

Low-value raw materials | Skilled migrants | Interest paid on loans | Loans | Small amounts of aid | High-value manufactured goods

Developing countries

Figure 2.4 Frank's dependency theory.

Exam tip

It's probably faster to sketch and label Rostow's model in the exam than trying to describe it in words.

Now test yourself

Which stage in Rostow's model are emerging countries at?

TESTED

Developing countries provide developed countries with cheap raw materials (cash crops, metal ores), skilled workers and interest on loans they took out to try and develop. Developing countries depend on the developed world for costly manufactured goods, aid and loans (which lead to debt).

The two theories are linked; the least developed countries are often trapped in dependency but if they can begin to modernise and develop infrastructure they can break free and develop independently, like China.

Dependency theory is rather simplistic. For instance, countries like China have broken free of this model and industrialised. However, some developing countries do seem to be too dependent on the developed world.

Exam practice

1 Define the term 'colonialism'. [2]
2 Explain how Rostow's modernisation theory can help understanding of the process of development. [4]
3 Assess the importance of historical, economic and political factors in explaining global inequality. [8]

ONLINE

Globalisation

Emerging countries have managed to break out of stage 2 of Rostow's model and develop. **Globalisation** is a major reason for this. Globalisation has caused investment by **transnational corporations (TNCs)** to spread around the world because:

- the internet and mobile phones have made communication and data transfer cheaper, faster and easier than ever before
- people travel more easily and cheaply around the world on fast jet aircraft
- goods can be moved cheaply on container ships.

These developments in technology have encouraged many TNCs from developed countries to invest in emerging countries in two main ways:

- shifting the production of goods like mobile phones, computers, TVs and clothes to emerging countries in Asia, especially China
- moving data processing, call centres and administrative functions abroad, especially to India.

By moving some of their businesses abroad, TNCs have created new jobs in emerging countries, meaning higher incomes and a better quality of life for some people.

Many Asian country governments have made it easy for TNCs to locate in their countries by:

- creating special economic zones (SEZs) and export processing zones (EPZs) for TNCs to locate in, which have no, or very low taxes
- restricting workers' rights, banning workers from joining unions and having no, or a very low, minimum wage
- having limited environmental, pollution, and health and safety laws, which reduces costs for new factories and offices.

Globalisation has benefited different countries, and groups of people, in very different ways (see Table 2.6).

Table 2.6 Impacts of globalisation in different countries

USA: male car factory workers	China: female factory workers
Companies like Ford and General Motors have shut car factories in cities like Detroit in the USA, and moved them to Mexico and Brazil Hundreds of thousands of male, well-paid workers lost their jobs as the jobs moved overseas to cheaper locations	Women who work for Foxconn, the company that makes iPads and iPhones for Apple, earn about £180 per month Most live in dormitories within the factory and work up to 60 hours of overtime a month, but pay is much higher than in the rural areas from which most workers migrated
Bangladesh: female textile workers	**Sub-Saharan Africa**
Many low-priced clothes are made in Bangladesh by women working for up to 80 hours a week but only earning as little as £12 Accidents, factory fires and exploitation are common – all in the name of cheap clothing	So far, globalisation has had very little impact here. The region has poor infrastructure, power and water supplies Education and skills levels are low, and governments are often corrupt. This has prevented large-scale investment by TNCs

Globalisation Growth in world trade, transport, migration and communication meaning the world is more connected than ever before

Transnational corporations (TNCs) Companies with operations (factories, offices, warehouses) in more than one country

Revision activity

Draw a spider diagram of all of the factors that have made Asia attractive to investment from TNCs.

Exam tip

Remember globalisation has both positive and negative impacts, and you need to be able to discuss both.

Now test yourself

Why have many Asian countries proved attractive places for TNCs to locate some of their business to?

TESTED ☐

Approaches to development

REVISED

Development strategies are used by countries or regions that want to develop. Usually this means implementing a project that aims to improve aspects of people's lives.

Top-down versus bottom-up

Development strategies can broadly be divided into top-down and bottom-up, which have very different characteristics (see Table 2.7).

Table 2.7 Top-down and bottom-up development

	Top-down development	Bottom-up development
Aims	Economic development: large projects designed to improve incomes for many people, often by developing industry	Social development: smaller projects that aim to improve health education or food supply at a local level
Scale	Large: such as a whole region or a city	Small: a village, a small rural area or an urban slum
Control	National: organised and controlled by central government in the capital city	Local: organised and controlled by the local community
Funding	Cost millions or billions of pounds and are sometimes financed by foreign loans and **IGOs** such as the World Bank	Very low cost (hundreds or a few thousand pounds), often funded by outside **NGOs** such as Oxfam or Practical Action
Technology	Often highly technical, using imported machinery and foreign technical support	Often this is **intermediate technology**, which is simpler and needs less technical support, or renewable energy technologies
Examples	Large hydroelectric power dams Major roads, bridges and railways New ports and airports	Wells and water pumps Schools and health clinics Training for farmers

Bottom-up approaches

A bottom-up example is the British NGO WaterAid which installs wells and hand pumps in Africa. These facilities only cost £292 each, and use intermediate technology that can be maintained and repaired by local people. WaterAid helps to provide clean water for a village, improve health and reduce the time women and children spend collecting water. However, thousands of wells are needed across Africa, so progress is slow for getting clean water to everyone. Clean water on its own does not improve people's incomes. Bottom-up development often meets basic needs, like health, water and food supply – but it may not increase people's incomes by very much.

Exam tip

Don't fall into the 'bottom-up good' 'top-down bad' trap; both approaches have pros and cons.

Now test yourself

TESTED

What is meant by intermediate technology?

IGOs (inter-governmental organisations) Regional or global organisations that bring governments together, such as the United Nations or the World Bank

NGOs (non-governmental organisations) Charities that often help low-income communities to develop by funding projects

Intermediate technology Simple technology that is easy to build and repair, suitable for the developing world and does not rely on outside help

Top-down approaches

Large-scale infrastructure schemes in developing and emerging countries are top-down and planned by governments. They are paid for by a combination of:

- loans from commercial banks and IGOs, such as the World Bank
- government investment
- investment by TNCs.

These multi-billion dollar projects are often controversial. Examples include the Addis Adaba to Djibouti railway line, completed in 2016 and costing $3.2 billion. This railway connecting land-locked Ethiopia to the coast at Djibouti was funded by the Export–Import Bank of China and built by the Chinese TNCs China Civil Engineering Construction Corporation and China Railway Group. China's own Three Gorges Dam is another example. This cost $26 billion, took fourteen years to build and created a 600 km long reservoir behind the dam wall. It generates 22,500 MW of electricity.

However, building the Three Gorges has had many impacts:

- 1.3 million people had to relocate to make way for the reservoir
- 1300 archaeological sites were flooded
- species like the Chinese river dolphin and Siberian crane are threatened by the dam
- the reservoir may become polluted with farm and industrial waste
- rice farmers below the dam no longer benefit from annual floodwaters irrigating their fields
- the lake will silt up in 50 years and flood control will get more difficult.

Top-down development projects, and investment by TNCs, often benefit urban areas and industry much more than rural areas and farmers. They can help countries to become more economically developed but might also contribute to greater inequality.

Now test yourself

TESTED

Which people often benefit most from top-down development?

Exam practice

1 Explain the term 'bottom-up strategy'. [2]
2 Explain why top-down development strategies often have costs. [4]
3 Assess the impact of globalisation on different countries. [8]

ONLINE

How is one emerging country managing to develop?

Case study: emerging China

REVISED

China's place in the world

China is an emerging country which is increasingly important globally. It is the world's largest country, with a population of 1.38 billion people in 2016. China is situated in the centre of Asia (see Figure 2.5) and shares international borders with fourteen other countries:

- China has a huge coastline of about 14,500 km, giving it excellent access to the sea for trade.
- China's vast interior extends far into central Asia.
- Major rivers, like the Mekong, Yangtze, Pearl and Yellow, extend far inland and in many cases can be navigated, allowing trade far into China.

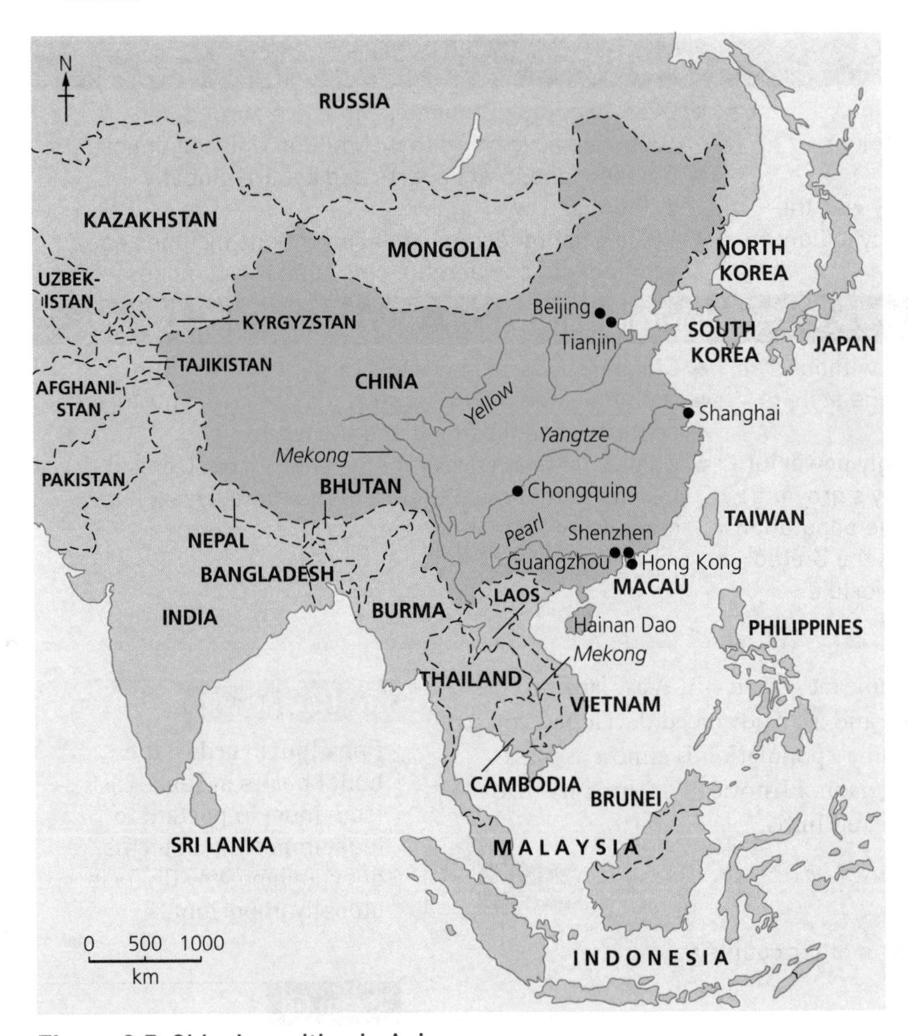

Figure 2.5 China's position in Asia.

China's situation means that it is close to all of the major trading nations in Asia and has good **connectivity**. Asia has a population of 4.5 billion people, or 60 per cent of the world's population. Japan, South Korea, China, Russia and India are all in the list of the world's twelve largest economies:

- In 2014, China had eight of the world's ten busiest container ports.
- Beijing, Hong Kong and Shanghai were all in the world's top ten busiest airports in 2016.
- In 2016, China had 22,000 km of high-speed railway, more than the rest of the world put together.

In other words, China has very good connections within the country, and to the rest of the world.

Connectivity The links between places, such as roads, railways, shipping routes and air travel, as well as internet connections

Exam tip

Your emerging (or developing) country case study requires detailed factual knowledge and data.

Now test yourself

TESTED

1 What was China's population in 2016?

China is significant in other ways (see Table 2.8).

Table 2.8 The significance of China

Socially	Culturally
• The world's most populous country • Chinese is the world's most spoken language • The 1979–2015 One Child Policy was the largest population control policy in human history	• Over 50 million Chinese people live abroad; ten countries have more than 1 million Chinese emigrants • Chinese food, martial arts and art are globally influential • China's list of technological inventions includes paper, printing, gunpowder, the clock and the compass
Politically	**Environmentally**
• China is a Communist country, without free elections, but with its own unique style of one-party government • Chinese leaders are increasingly powerful globally, because of the country's growing wealth, military power and huge population • China has a permanent seat on the United Nations Security Council, the world's ultimate decision-making body	• China is, by far, the world's largest CO_2 emitter • Pollution, and how to control it, is a key twenty-first century issue for China and the world • China's natural environment ranges from tropical forests to mountain tundra in the Himalayas – it has globally important biodiversity

China is not alone in being an important country in Asia. Japan is smaller, but much richer per capita and is an advanced developed country as well as a key ally of the USA. India's population is almost as big as China's but it is much poorer per person. Historically, China has had tense relationships with both Japan and India.

Revision activity

Rank (put in order) the bullet points in Table 2.8 from most important to least important in terms of explaining why China is globally important.

Now test yourself

TESTED

2 How many Chinese people live in other countries?

Exam practice

1 Using Figure 2.5, describe China's location within Asia. [2]

2 Suggest two ways in which China can be considered a globally significant country. [4]

ONLINE

Exam tip

China is a powerful, wealthy country but it is not yet a developed country like the UK or USA. Students often claim it is!

China's economy

REVISED

China's economic development since 1990 has been phenomenal. In 1990, China was only the world's eleventh largest economy by GDP whereas today it is the second largest. Figure 2.6 shows how China's GDP has grown since 1990.

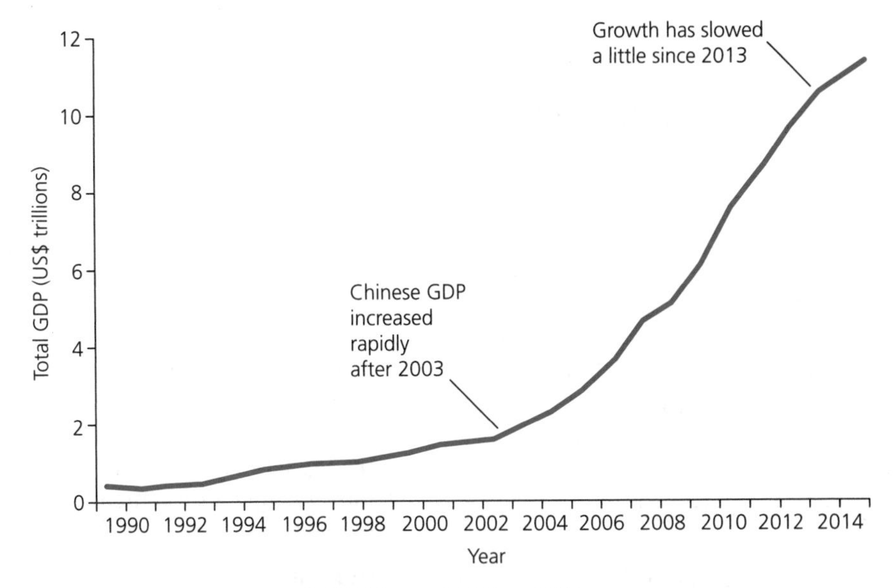

Figure 2.6 Chinese total GDP from 1990 to 2014.

Perhaps more important than the growth in GDP is the growth in per capita **gross national income (GNI)**. The data in Table 2.9 shows that the average Chinese has moved from a very low, developing country, income per person in 1990 to a middle-income level by 2015. In other words, China changed category from the developing to the emerging world within a generation. The average Chinese person earned 25 times more in 2015 than they did in 1990.

Gross national income (GNI) All of the income earned by a country in a year from within a country and any money earned abroad

Economic sectors The broad divisions of an economy into primary, secondary and tertiary jobs

Table 2.9 China's growth in per capita GNI from 1990 to 2015

China	1990	1995	2000	2005	2010	2015
GNI per capita (US$)	318	610	960	1755	4560	8027

Revision activity

Use the data on Chinese GNI. Draw a line graph to show how China's GNI has changed.

A huge shift in **economic sectors** has taken place (Figure 2.7) as primary industry has shrunk as a share of the economy and jobs have moved into manufacturing and services. This has been accompanied by a huge movement of people from living in the countryside to living and working in cities.

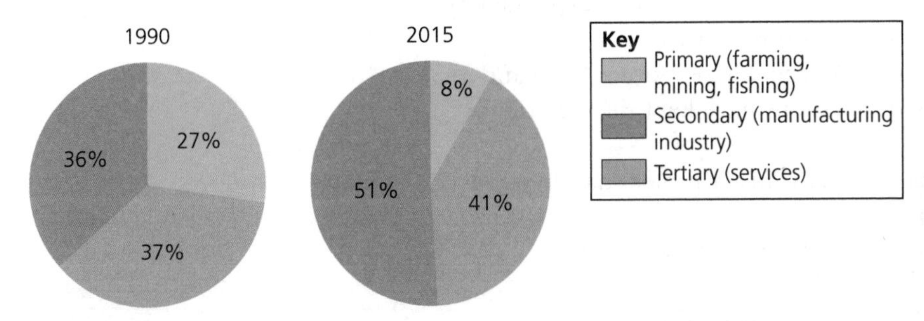

Figure 2.7 Chinese economic sectors in 1990 and 2015 (percentage of total GDP).

Exam tip

The data in Figure 2.7 is the sort of data you need to commit to memory and use in the exam.

China is often referred to as the 'workshop to the world' because so many products we buy are made in China. This can be seen in China's main imports and exports. Its top imports in 2015 were mostly raw materials or components (crude oil, iron ore, gold and printed circuit boards) but its top exports were all manufactured goods (computers, mobile phones, printers and so on).

Now test yourself TESTED

How did China's economic sectors change between 1990 and 2015?

Globalisation and China

Of all the world's emerging economies, China has benefited the most from globalisation. There are several reasons for this:

- In the 1990s, **containerisation** and the rise of the internet meant that developed world TNCs could easily move factories to China (called **outsourcing**) and ship goods to Europe and North America, but still keep in touch using the internet to track orders and shipments.
- China had a huge pool of low-cost, well-educated and skilled labour.
- An education law in 1986 provided free primary and secondary school education for everyone.
- The Chinese government had begun its 'Open Door Policy' in 1980, which was strongly in favour of **foreign direct investment (FDI)**.
- This policy encouraged foreign TNCs to invest in China's EPZs and SEZs.

Containerisation Transporting goods in standardised metal shipping containers, which can be transported cheaply (and increasingly automatically) by train, truck or ship

Outsourcing TNCs shifting some of their factories and/or offices overseas to cheaper locations

Foreign direct investment (FDI) Money from one country invested into another, for example a TNC from the USA building a factory in China

FDI flooded to China, especially after the year 2000. The main sources of FDI into China in 2012 are shown in Table 2.10.

Table 2.10 Main sources of FDI into China in 2012

Country/region	Amount (billions of US$)
Europe	3.5
North America	2.6
Japan	7.3
Hong Kong	65.6

Hong Kong, which is a semi-independent province of China and an important financial centre, acts as a 'middle man' for Chinese FDI. Much of the Hong Kong FDI is actually from American, Japanese and European TNCs that use banks in Hong Kong to channel their money into China.

China's long period of rapid economic growth can be partly explained by the government's relentless investment in infrastructure:

- Huge hydroelectric power schemes like the Three Gorges Dam and hundreds of coal and nuclear power stations have kept the electricity flowing.
- The huge $70 billion south–north water transfer project has diverted water from the wet south to the dry north.
- 22,000 km of high-speed railways have been built, with over 30,000 km more planned.
- Since 1990 China has built 123,000 km of major roads.

An unusual feature for an emerging Asian country is that China has never relied on **tied aid** or **multi-lateral aid** to help it develop. Countries like South Korea, Thailand and Indonesia have been helped to develop by aid from some developed countries, especially the USA and Japan. China has done it differently, by attracting foreign TNCs and reinvesting the money earned in infrastructure and education.

Exam tip

It's worthwhile learning definitions of complex terms like 'globalisation' because they are hard to define off the top of your head.

Exam practice

1 Describe the trend in Chinese GDP shown in Figure 2.6. [2]
2 Explain how the economic sectors of the Chinese economy have changed over time. [4]
3 Define the term 'foreign direct investment'. [2]
4 Explain how the government of China has helped the process of globalisation. [4]

ONLINE

Now test yourself

Which countries and regions are the main sources of FDI flowing into China?

TESTED

Changing China

REVISED

In emerging countries, development often leads to demographic change:

- As people move to cities and get jobs in factories and offices the fertility rate falls.
- Many workers in TNC factories are young women; this reduces the fertility rate further as the women marry later and may focus on their careers.
- Fewer people work on farms, so there is less need to have children to help bring in the crops.
- As a country gets wealthier, healthcare improves and death rates drop because people are healthier and live longer.

China did things differently. Faced with a rapidly growing population in the 1970s, and fears that food and water would run out, China introduced the One Child Policy in 1979. This forced down the fertility rate by government edict, rather than it falling naturally due to economic development.

Development has led to massive **urbanisation** in China (see Figure 2.8, page 42). Between 1990 and 2015, about 300 million people moved from the countryside to cities in a process called **rural–urban migration**. These are often people aged 18–30 seeking jobs in the factories and offices of China's rapidly growing cities.

China's cities have seen phenomenal growth and many are now **megacities** (see Table 2.11).

Table 2.11 The growth of some of China's cities

	Population (millions)	
City	**1990**	**2015**
Guangzhou	5.9	20.8
Shanghai	13.3	24.5
Chongqing	14.0	29.9
Beijing	10.8	21.5

Tied aid Money, or technical help, from one country to another that has 'strings attached', so something is expected in return

Multi-lateral aid Money, or technical help, given to developing or emerging countries by an IGO like the World Bank

Urbanisation The increase in the proportion of a population living in towns and cities

Rural–urban migration Migration of people from the countryside to towns and cities

Megacity A city with a population of over 10 million people

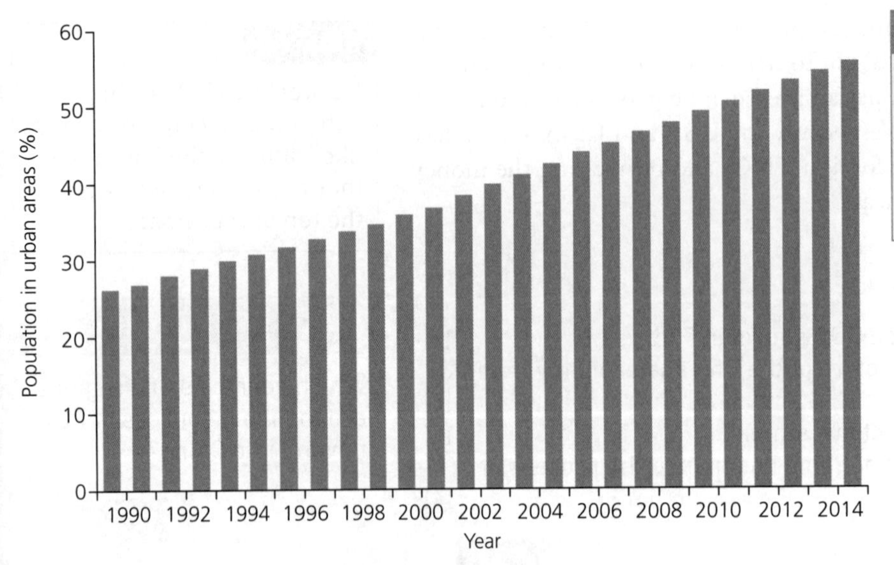

Figure 2.8 China's urbanisation level from 1990 to 2014.

Revision activity

Make a list of reasons why young people might want to leave rural China, and a list of things that pull them towards cities.

Now test yourself

TESTED

Why has China's population changed in a unique way, compared to other emerging countries?

Exam tip

In the exam, you can't just refer to 'China'; you will be expected to name some of its cities and provinces.

Positive and negative impacts

China's transformation into an emerging economy has been uneven. Figure 2.9 shows how income per person varies across China. China's coastal cities have been most affected by urbanisation, in-migration and globalisation. They are the economic **core areas** of China. China's rural interior has been changed much less and suffered out-migration of young people. These regions are China's economic **periphery areas**.

Core areas Urban places where most industry, and therefore most wealth, is located

Periphery areas Isolated and usually rural places. People tend to leave these areas and migrate to the cores

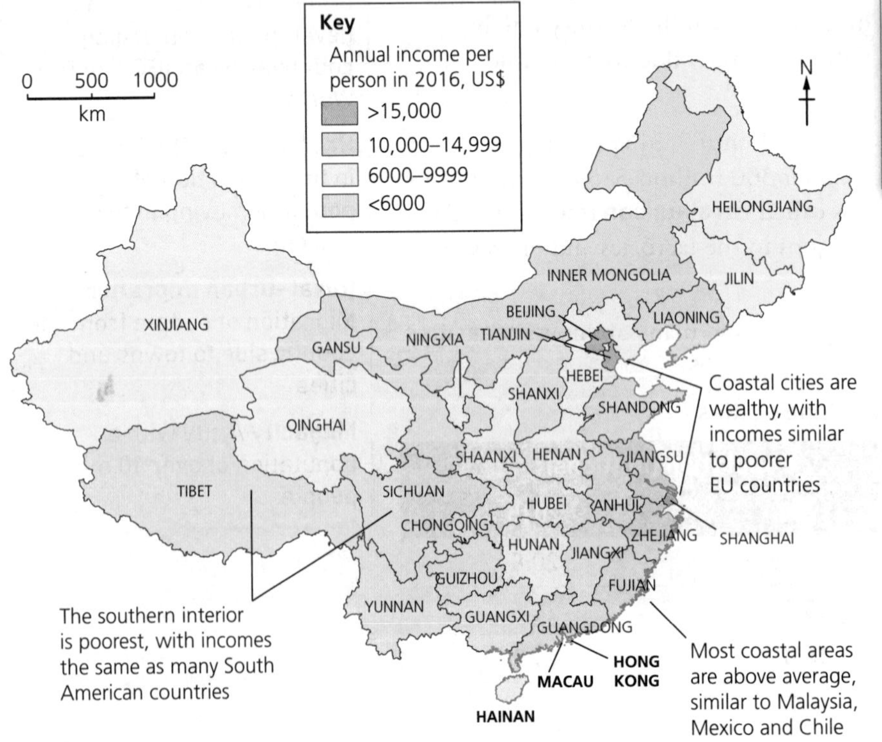

Figure 2.9 Income variation across China.

Comparing two Chinese provinces reveals **socio-economic** differences (see Table 2.12).

Socio-economic
Characteristics of groups of people such as income, age, gender, education and health

Table 2.12 Socio-economic differences between two Chinese provinces

Factor	Sichuan (58% rural population)	Shanghai (89% urban population)
Life expectancy (years)	75	80
Population aged over 65 (%)	13.9%	9.6%
Population growth 2000–10	–2.3%	+40%

- In Sichuan, many rural people still live in poverty. Sichuan has seen its population shrink as people have migrated to cities. The people live in traditional villages, but these suffer from a lack of young people due to rural–urban migration. Young people living in cities often visit their families during Chinese New Year but then return to Shanghai or Beijing.
- In cities like Shanghai incomes are higher but housing and the cost of living are expensive. Chinese young people work long hours in offices and factories. Finding a partner can be hard because of the impact of the One Child Policy. There are far more men than women, and overall few young people.

Many families in China face the '4–2–1 problem' because of the One Child Policy. One child has to look after 2 parents and 4 grandparents. Often the child has moved to a city, but the parents and grandparents live in a rural area.

Now test yourself

Explain why Sichuan's population fell between 2000 and 2010.

TESTED

Impacts on different groups of people

China's development has had many impacts on different groups of people, in terms of both age and gender (see Table 2.13).

Table 2.13 The impact of China's development on different groups of people

Female factory worker, age 16–30	Rural male farmer, age over 50
Many Chinese rural–urban migrants work in factories making consumer goods for export. A 60-hour week is quite normal, and pay is around \$150–300 per month. These migrants have left their families behind	In some ways, rural farmers are the forgotten people of China. Farming is hard, labour-intensive work and some farmers earn less than \$1000 per year. Farmland close to cities is often seized for new housing development and roads
However, living conditions, access to services and opportunities are better than in many rural areas, so overall people may be better off	Even the countryside of China is increasingly polluted by industry and mining, and water supply is often poor
Urban businessman, age 30–50	**Female office worker, age 30–50**
It's estimated that Shanghai has over 150,000 millionaires. Globalisation has created business opportunities and many urban people have become very wealthy	Educated women in China have increasingly entered the workforce and earn good salaries (\$15,000–25,000 per year) in retail and professional services
However, Chinese cities are some of the most polluted and congested in the world, so is quality of life really that high?	However, there is a still a long way to go on gender equality, and women's education and pay are still not as good as men's

- The main question in China is whether living and working in one of China's coastal cities is better than life farming in the countryside. Over the past 30 years, about 25 per cent of China's people have switched from the countryside to cities.
- As in most emerging countries, people are feeling the benefits of higher pay and greater opportunities in cities but also experiencing costs. Polluted, congested and expensive cities are often not great places to live.
- Over the next 30 years, Chinese cities are likely to improve and become more like developed world cities, that is they will be cleaner and better run. In the meantime, life in emerging world cities is often hard.

Now test yourself

Which type of people have the lowest incomes in China?

TESTED

Exam practice

1 Define the term 'rural–urban migration'. [2]
2 Explain two reasons why some parts of China have higher per capita incomes than others. [4]
3 Assess the positive and negative impacts of globalisation on different groups of people. [8]

ONLINE

Exam tip

If you get a question about different groups of people, try to be a bit more specific than 'young and old' or 'urban and rural'.

Environmental impacts

China's rise as an emerging country has come at a huge environmental cost. In 2005, China became the world's largest emitter of carbon dioxide (CO_2), the **greenhouse gas** that causes global warming (Figure 2.10). China emits almost 30 per cent of all global CO_2. This means:

- China is the largest contributor to global warming
- any global agreement to reduce greenhouse emissions would have to include China.

Greenhouse gases Gases thought to cause the enhanced greenhouse effect and global warming, namely methane, carbon dioxide and nitrogen oxides

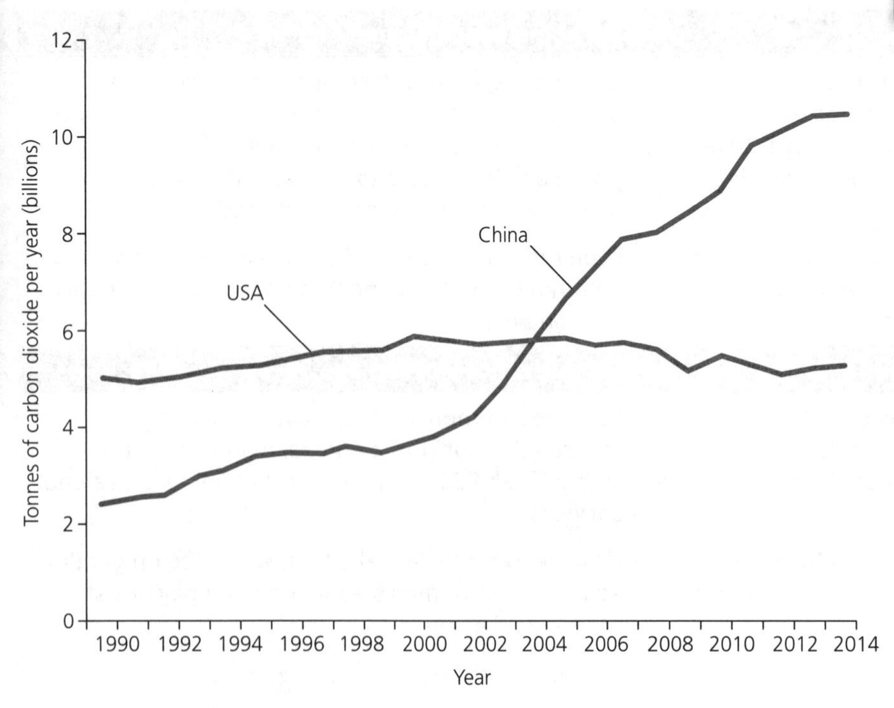

Figure 2.10 Carbon dioxide emissions from China and the USA.

China's emissions come from factories that make the consumer products we buy (computers, iPhones, TVs) and from its many coal-fired power stations. China uses dirty coal for 73 per cent of its electricity generation.

Coal-fired power stations, factories and an increasing number of cars contribute to China's other major **pollution** problem – air quality in cities:

- Ninety-nine per cent of China's city dwellers breathe air that is polluted.
- Sixteen of the world's twenty most polluted cities are in China.
- Air pollution and smog is sometimes so bad that schools and even airports are forced to close.
- Estimates suggest that about 1.5 million Chinese people die every year from health problems caused by polluted air.

Pollution Chemicals or other substances that are present in the environment (air, water, soil) because of human activity, that would not be there naturally

There are other environmental problems:

- Deforestation and desertification are widespread, as land in the countryside is overexploited.
- Farmland and soil are often contaminated with heavy metals from industry.
- Groundwater in 60 per cent of Chinese cities is of 'bad' or 'very bad' quality.
- Twenty-five per cent of Chinese rivers have water quality that is 'unfit for human contact'.
- Car ownership grew from 27 million in 2004 to 154 million by 2015, contributing to increased traffic deaths, congestion and poor air quality.

Now test yourself

TESTED

In which year did China become the world's largest emitter of carbon dioxide?

Revision activity

Draw a spider diagram showing all of China's environmental problems.

Exam practice

Explain two environmental problems caused by rapid economic development. [4]

ONLINE

Exam tip

The data on pollution on this page is very useful to learn, as it allows you to show how serious the problems are using statistical evidence.

China's influence

REVISED

China's economic development has changed its **geopolitical influence**. In 1990, as the world's eleventh largest economy China, was not that important compared to other countries. By 2016:

- China was the world's second largest economy
- it accounted for about fifteen per cent of the world's GDP.

This means that China cannot be ignored. If its economy grows, it helps other economies around the world to grow. Within Asia, China is now the largest economy, having overtaken Japan.

China has invested some of its wealth in becoming the largest military power in Asia, making it the strongest regional power (see Table 2.14).

Geopolitical influence
The power of a country as a result of its economic wealth, military power, technology and human resources

International relations
The extent to which countries get on with each other and cooperate, or don't

Table 2.14 Military power in Asia in 2016

Country	Tanks	Fighter aircraft	Destroyers	Nuclear weapons?
China	9150	1230	32	Yes
India	6464	679	10	Yes
South Korea	2381	406	12	No
Japan	678	287	43	No

China is so strong militarily and economically that it sits at the 'top table' in world affairs:

- China has a permanent seat on the United Nations Security Council.
- It is an influential member of the World Bank and International Monetary Fund.
- It agreed to reduce its CO_2 emissions after 2030, at the global climate summit in Paris in 2015 – something it had never agreed to before.

Despite its power and influence, China's **international relations** with other countries are often strained:

- European countries often criticise China because of its poor record on human rights; as a Communist dictatorship there are no free elections, the internet is censored and there is no freedom of speech.
- The USA is an ally of both Japan and South Korea, and these two countries have strained relations with China which means the USA usually sides with those countries.
- The USA has often accused China of 'stealing US jobs' because many US companies have relocated factories to China.
- China claims many small islands in the South and East China Seas which are also claimed by South Korea, Taiwan, Malaysia, Japan and the Philippines. The disputed ownership of these islands causes regional tensions.

In Asia, it is difficult for other countries to oppose China now, because it is so powerful.

Now test yourself

TESTED

Which other Asian country is closest to China in terms of military strength?

Costs and benefits

China is expected to have an economy similar in size to that of the USA by 2030 (see Figure 2.11). By then, it will probably have a per capita annual income of about $20,000–25,000. This would mean China will still be an emerging economy, but it will be much closer to being a developed one.

Past and future growth in China has a number of different economic and political costs and benefits (see Table 2.15).

Table 2.15 Growth in China: costs and benefits

	Economic	Political
Costs	Some people argue that China has gained jobs from countries like the USA and UK, so the developed world has suffered As China develops, wages rise and this could mean TNCs start moving to lower cost places like Vietnam and the Philippines – meaning jobs losses Some TNCs, like Apple and Nike, have been criticised over the poor working conditions in their Chinese factories	China could 'bully' its Asian neighbours as it becomes an even more powerful regional leader The USA may begin to see China as a direct threat to its position as the world's most powerful country A more powerful China may feel it can ignore calls to be more open and democratic and to respect human rights
Benefits	FDI by TNCs has created, and will continue to create jobs in China China has moved 500–600 million people out of poverty since 1990 and this should continue Over time, incomes and quality of life should improve even more Consumers in developed countries have benefited from cheap goods imported from China	China could become a leader on global issues such as pollution and global warming; it is already the world's leading user of wind and solar power China is already investing in Africa, where Chinese companies are creating jobs and building infrastructure – helping the world's poorest continent

China has two huge challenges to face by 2030:

- cleaning up its environment and reducing pollution so people can enjoy a high quality of life, rather than just getting wealthier
- meeting the resource demands of its 1.4 billion population in terms of water, food, consumer goods, housing and health services as people get wealthier and demand more and more.

Now test yourself

TESTED

Will the average Chinese person be as wealthy as the average American by 2030?

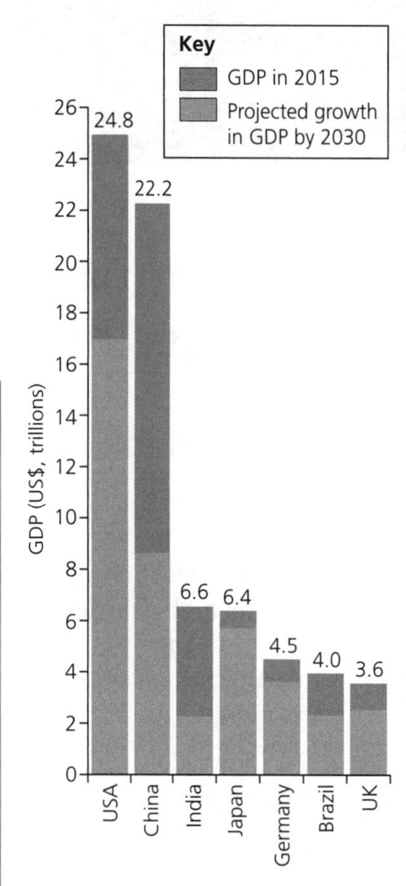

Figure 2.11 Total GDP for six countries in 2015 and projected to 2030.

Exam tip

Questions that ask about 'costs and benefits' or 'advantages and disadvantages' need balanced answers where you cover both, equally.

Exam practice

Assess how far economic development in China has changed its relationship with the rest of the world. [8]

ONLINE

Topic 3 Challenges of an urbanising world

What are the causes and challenges of rapid urban change?

Urbanisation on the rise

REVISED

Since 2007 over 50 per cent of the world's population has lived in towns and cities:

- Levels of **urbanisation** are linked to levels of development.
- Usually, the higher the level of economic development, the more urbanised a country is.
- This means that emerging countries are experiencing rapid urbanisation today.

Figure 3.1 shows global and regional urbanisation levels since 1980 and **projections** to 2050.

Urbanisation An increasing proportion of people living in cities, rather than the countryside or rural areas

Projections Best guesses of future size or the value of something

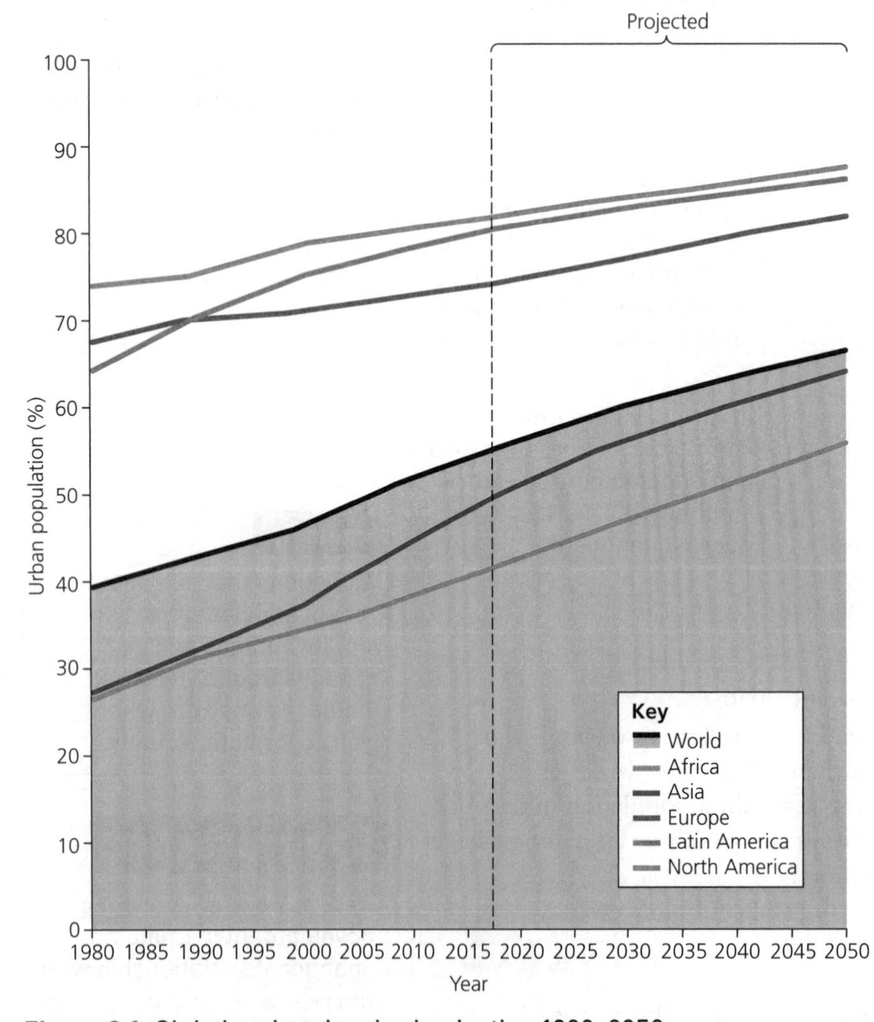

Figure 3.1 Global and regional urbanisation 1980–2050.

Notice from Figure 3.1 that:

- Developed regions (North America, Europe) have high urbanisation and future increases are small.
- Latin American urbanisation is high: it increased rapidly from 1980 to 2010 but has slowed since.
- Africa was only 42 per cent urban in 2017, but is expected to become rapidly more urban, reaching 55 per cent by 2050.
- Asia's urban population grew rapidly from 1995 to 2017, and this is projected to continue.

Developed countries are 'fully urbanised' because ten to twenty per cent of people will always live in rural areas. In emerging and developing countries, much more urbanisation is expected. There were 3.9 billion urban people in 2015. This is expected to increase to 6.3 billion by 2050 – equal to what was the entire world's population in 2003.

Revision activity

Roughly colour in a blank world map with regions of over and under 65 per cent urbanisation.

Exam tip

Make sure you can describe trends and rates of growth (line steepness) on a graph like Figure 3.1.

Now test yourself

TESTED

Which is the least urbanised continent?

Megacities

A key trend in urban growth is the rise of megacities. These exist on all continents but most are in Asia, especially the largest megacities with populations of over 23 million:

- Developed world megacities (London, New York, Tokyo) are growing slowly at about one per cent a year.
- Emerging megacities in China, Brazil and Mexico are growing quickly (two to three per cent a year), but growth rates are slowing.
- In low-income developing countries growth is very rapid, at over three per cent a year in cities like Lagos in Nigeria.

At a population growth rate of three per cent a year, a city will double in size in 23 years. Figure 3.2 shows the global distribution of megacities in 2017, when there were 38. This number will increase to around 45 by 2030.

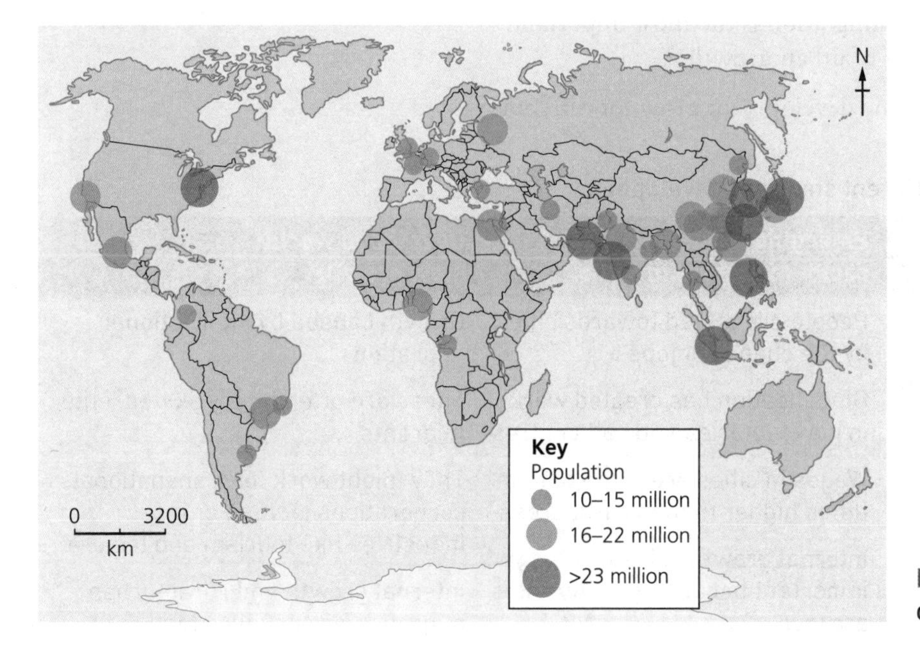

Figure 3.2 The global pattern of megacities in 2017.

Some countries have **urban primacy**: the largest city is a primate city and the next largest city is much smaller, perhaps only one-fifth the size (see Table 3.1).

Table 3.1 Urban primacy

Country	Primate city (population millions)	Next biggest city (population millions)
UK	London (13.8)	Greater Manchester (2.5)
Mexico	Mexico City (22.2)	Guadalajara (4.4)
Nigeria	Lagos (21)	Kano (3.5)

Primacy can cause geographical issues:

- People migrating to urban areas for work go only to the primate city.
- New businesses and investment are drawn to the primate city: smaller cities miss out.
- Rapid primate city growth means concentrated air pollution, poverty, traffic congestion and crime.

Mexico City generates 30 per cent of the gross domestic product (GDP) of Mexico, London 22 per cent of the UK's GDP. This gives these cities disproportionate economic and political influence:

- Economic decisions like spending money on infrastructure (roads, airports) will be made to benefit the primate city.
- Political power, that is national government, is often located in the primate city and its decisions reflect the needs of the city more than the needs of smaller cities and regions.

Urban primacy In some countries the largest city is much bigger than the next biggest, and is economically dominant

Now test yourself

Name a primate city.

TESTED

Exam practice

1 Calculate the change in percentage urbanisation for Asia between 1980 and 2017 shown in Figure 3.1. [2]

2 Describe the global pattern of megacities shown in Figure 3.2. [4]

3 Explain one problem caused by urban primacy. [2]

ONLINE

Economic and social change

REVISED

Urbanisation happens in three ways:

- People migrate to cities (called rural–urban **migration**).
- Babies are born in cities (called internal growth).
- Villages and small towns are reclassified as 'urban' areas, not rural settlements.

Migration Where people move to a new location to live for more than one year

In developing and emerging cities, migration is the most important process, accounting for 60 per cent of urban growth.

Cities at different stages of economic development grow for different reasons (see Table 3.2).

Table 3.2 Growth in cities at different stages of development

Developing world: very rapid growth	Emerging world: rapid–moderate growth	Developed world: slow growth
Rural–urban migration caused by poverty, desertification, conflict and poor water supplies in rural areas People are forced to move to cities to survive Fertility rates are high so internal growth is also rapid	People are pulled towards cities by the chance of jobs Globalisation has created work in new factories and call centres Wages in cities are three to four times higher than in rural areas Internal growth is less important because fertility rates are low	Growth caused by international migration These are often highly skilled, elite migrants They might work for transnational corporations (TNCs), or in industries like tourism and leisure Internal growth and rural–urban migration are usually small

Some developed world cities have seen population decline. These include Detroit (USA), Leipzig (Germany) and Glasgow and Liverpool (UK). The decline has been caused by **deindustrialisation**:

- Old factories from the Industrial Revolution in the nineteenth century closed in the 1970s because they were old and inefficient.
- Cities were built near natural resources like coal and iron ore: once these were exhausted the factories closed.
- TNCs have closed high-cost factories and relocated them to cheaper locations in China and Mexico, as part of globalisation.

Once cities start weakening, they can enter a spiral of decline: closed factories, derelict land, crime and vandalism make them unattractive, so more people and businesses leave.

Deindustrialisation The loss of jobs in the secondary economic sector (manufacturing industry) as factories close

Working conditions Include pay rates, health and safety at work, hours, allowed breaks, and paid holidays and maternity leave

Revision activity

Make three revision flash cards to contrast the characteristics of developing, emerging and developed world cities.

Exam tip

You need to be clear which countries are developing, emerging and developed, so learn a short list of each.

Urban economies

The experience of working in cities is very different depending on whether work is formal or informal employment:

- Informal: no contract of employment, earnings vary from day to day, no tax is paid, poor working conditions, child labour is used, unsafe and unregulated **working conditions**.
- Formal: contract of employment, taxes are paid, regular weekly wage or monthly salary, employment laws are enforced, health and safety regulations are enforced.

In developing world cities, informal work accounts for most economic activity whereas in developed cities the reverse is true (see Table 3.3).

Table 3.3 Informal and formal employment in developing, emerging and developed countries

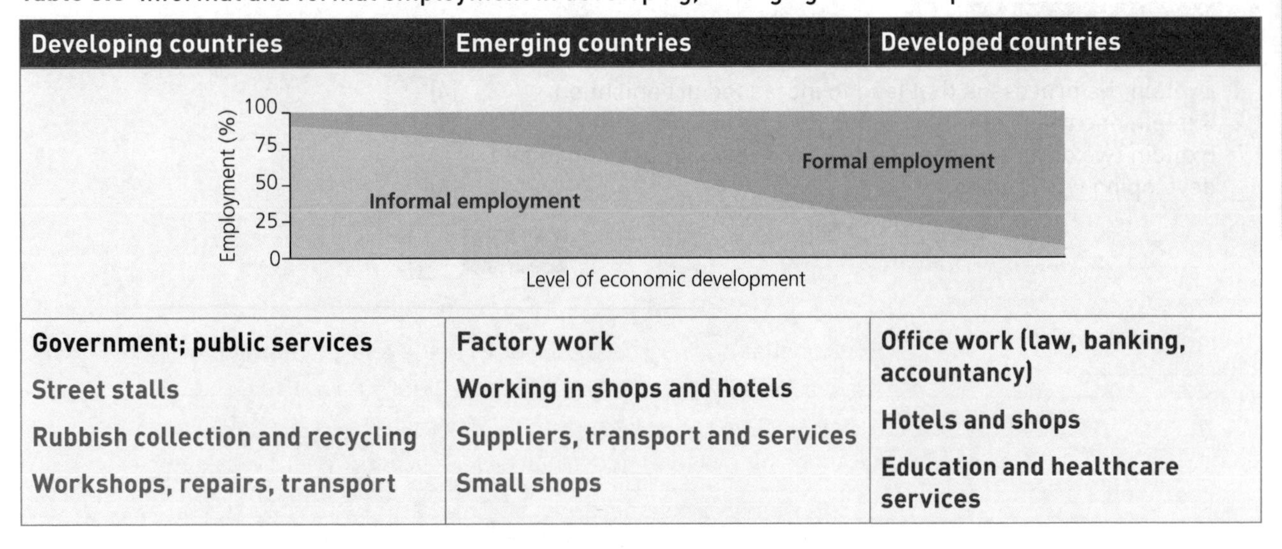

Developing countries	Emerging countries	Developed countries
Government; public services Street stalls Rubbish collection and recycling Workshops, repairs, transport	Factory work Working in shops and hotels Suppliers, transport and services Small shops	Office work (law, banking, accountancy) Hotels and shops Education and healthcare services

As cities become more developed, their economies become more formal. In a developed city, the only informal employment is illegal, such as employing undocumented migrants or 'cash in hand' work to avoid paying tax.

Economic sectors are also different in contrasting cities (see Table 3.4).

Table 3.4 Economic sectors in contrasting cities

City (percentage employment)	Primary sector	Secondary sector	Tertiary sector (services)		
			Financial and business	Transport and trade	Other
London (developed)	–	6%	33%	31%	30%
Shanghai (emerging)	7%	33%	13%	26%	21%
Lagos (developing)	10%	20%	8%	45%	17%

- London is dominated by the service sector, which accounts for 94 per cent of jobs. London is globally important for banks, insurance companies and accountancy firms, so many jobs are in business and finance and are highly paid.
- In the emerging world Chinese city of Shanghai, the largest number of workers is in secondary employment, because there are many factories making goods for export; wages are low but the work is regular, even though people work long hours in harsh conditions.
- Lagos in Nigeria is dominated by transport and trade employment; much of this is informal and involves moving goods and people around the city, small informal shops and street sellers.

Now test yourself

TESTED

Which types of cities have most employment in the informal sector?

Revision activity

Make a list of informal and formal jobs.

Exam practice

1 Explain the processes that lead to increased urbanisation. [4]
2 Explain one reason for population decline in some cities. [2]
3 Explain two differences in employment between developed and developing world cities. [4]

ONLINE

Changing cities

REVISED

Cities broadly follow a 'cycle of urbanisation' over time, shown in Figure 3.3.

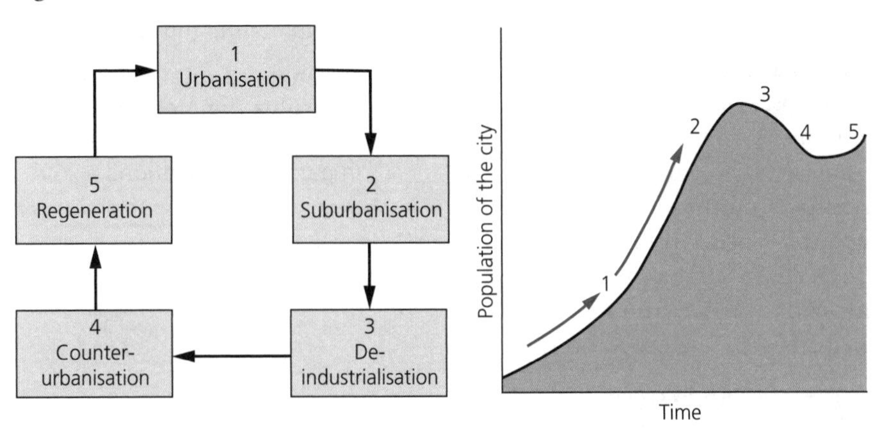

Figure 3.3 The cycle of urbanisation.

Table 3.5 The cycle of urbanisation explained

Stage	Place in cycle
1	Cities grow rapidly in both population and **spatial growth**, but tend to be high density In developing cities, this includes the growth of **slums** Rapidly urbanising cities are congested, polluted and chaotic
2	Wealthier people move to new, better, lower density housing on the city edge, leading to large spatial growth (called urban sprawl) and the development of suburbs Suburbanising people are trying to get away from the problems in the centre of the city
3	Many developed world cities have experienced deindustrialisation Factories have closed, creating abandoned and derelict land and run-down areas, making the city unattractive as well as reducing the number of jobs
4	Some people respond to the urban decline by leaving the city altogether and going to live in rural areas close by They commute into the city to work
5	In order to make cities attractive again, local and national government invest money to regenerate cities – especially their run-down, deindustrialised inner city areas Regeneration involves building new housing, better roads, business parks to attract companies back and leisure and recreation spaces like parks

Spatial growth The pattern of growth outwards as a city expands in area over time

Slums Informal housing built by their owners out of discarded materials, often on wasteland or as high-density apartment blocks. Slums lack water, electricity and sanitation

Now test yourself

TESTED

What is the movement of people out of urban areas into rural areas called?

Exam tip

Both the cycle and graph in Figure 3.3 could easily be sketched in the exam, saving you time.

Urban land use

All cities have different **land-use zones**. These include:

- Commercial land use: shops, offices, hotels and other businesses cluster together in the central business district (CBD) at the centre of the city. They can also be found along major roads, and in developed world cities in city-edge (out-of-town) retail and business parks.
- Industrial land use: factories, workshops, science parks and office complexes. Older manufacturing industry is often found close to the city centre in an area called the inner city. Newer industry is often found on purpose-built industrial estates on the city edge.
- Residential land use: this takes up most space in a city, and is the housing where people live. It is high density (flats, apartments) close to the city centre, but density falls with distance from the city centre. Suburbs have low-density detached and semi-detached housing.

Land-use zones Areas in cities that have particular functions, like retail (shops) or residential (housing)

Planning regulations Laws and rules that determine what can and cannot be built in particular locations

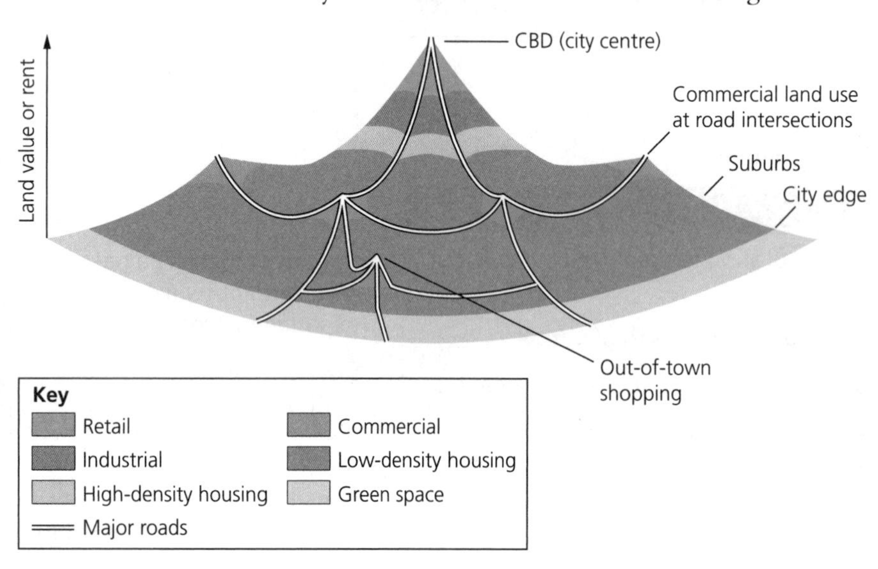

Figure 3.4 Explaining land-use patterns.

Land-use patterns can be explained by several factors (Figure 3.4):

- Cities have a small, very accessible area at their centre where roads and railways meet: the CBD. High accessibility and low availability (because the area is small) mean that large shops, hotels and offices want to locate here. This creates high land costs: too expensive for housing.
- Industry locates close to, but not in, the CBD. It needs good access but will not pay as much as large retail to buy or rent land.
- Commercial land uses often locate along major roads, ring-roads and large intersections because they get passing customer traffic.

Land use is also influenced by **planning regulations**:

- Cities are often surrounded by protected green space, called a greenbelt, that cannot be built on.
- In some cities, historic areas are protected as conservation zones, meaning old buildings cannot be altered or demolished.
- City governments might remove planning regulations in run-down areas, to encourage regeneration.

Now test yourself

TESTED

What type of land use is found in the CBD of cities?

Exam practice

1. Define the term 'suburbanisation'. [2]
2. Explain two reasons why land values are highest in the CBD of a city. [4]

ONLINE

Why does quality of life vary in megacities?

Case study: Karachi, Pakistan

REVISED

- Pakistan is a developing country. It had a human development index (HDI) score of 0.38 in 2016, meaning low human development.
- Karachi is Pakistan's largest city, with a population of about 24.3 million in 2016. The next biggest city is Lahore, with a population of 10 million.

Site The land the city is built on

Situation The position of the city in relation to surrounding areas

Connectivity The number of links such as road, rail and airlines connecting the city to other places as well as connections made by people through migration or TNCs

Megacity context

Karachi is a coastal city located on the Arabian Sea in southern Pakistan. It is the seventh largest megacity in the world. Both Karachi's **site** and **situation** explain why it has become a major megacity (see Table 3.6).

Table 3.6 Karachi's site and situation

Site	Situation
Flat coastal plain that provided farmland Coast gives access to the sea for fishing Coastal mangrove trees for fuelwood and building materials A natural harbour, giving shelter for fishing and trading boats Two small rivers, the Malir and Lyari, for freshwater supply	Nearby Indus River provides access into the interior of Pakistan Karachi's harbour allowed it to be a port on trading routes between the Middle East and South Asia The fertile land and fishing waters of the Indus Delta provided growing Karachi with its food supply, and goods to trade

Exam tip

Make sure you can locate Pakistan and Karachi on a world map.

Now test yourself

How many people live in Karachi?

TESTED

Most of the world's megacities are located at the coast. This shows how important trade is in the development of very large cities.

Karachi has a high degree of **connectivity** at different scales, and because of this it creates twenty per cent of Pakistan's total GDP (see Table 3.7).

Table 3.7 Karachi's connectivity

National	The M-9 motorway linking Karachi to Islamabad (the capital city), and the M-5 to Lahore are being built as part of the China–Pakistan Economic Corridor (CPEC) The mainline railway links Karachi to Lahore and Islamabad in the north Karachi dominates secondary sector employment in Pakistan, with about 40 per cent of all manufacturing located there
	Karachi is home to Pakistan's stock exchange, and all of its major banks
Regional	Karachi is at one end of the CPEC being developed by China and Pakistan, giving China an export route for the sea from its western regions Karachi, and nearby Gwadar, are important ports on the shipping route between Asia and Europe
Global	Karachi is the major hub for sub-sea fibre-optic internet cables, seven of which enter Pakistan there Jinnah International Airport is Pakistan's largest, with over 6 million passengers per year connecting to 70 other cities Global TNCs that have offices and factories in Pakistan are mostly located in Karachi, for example Toyota and Suzuki

Megacity structure

Karachi is typical of many developing world cities in term of its **urban structure** (see Figure 3.5):

- Its CBD is located close to the port area. This area contains some historic buildings that date from when Pakistan was a colony of the UK and part of British India (until 1947).
- The city has grown outwards and inland from this area.
- Industrial areas circle the port area of the inner city, and are found on the east and south banks of the Malir River: factories are located close to port transport facilities and in some cases along major roads leading into Karachi.
- Close to these industrial zones are areas of slum housing, called *katchi abadis* in Pakistan.
- Many low-income workers live in slum housing close to where they work in formal and informal factories and workshops.
- Formal housing is found stretching from the inner city through suburban areas to the city edge.
- Clifton is a wealthy suburb relatively close to the CBD: in developing world cities the wealthy often live in older, central areas whereas in developed world cities they tend to live in city-edge suburbs.
- Karachi's **urban–rural fringe** is an area where informal slum housing is often built, by rural–urban migrants who have just arrived.

Urban structure The pattern of land use in a city

Urban–rural fringe The area at the edge of a city where it meets the surrounding countryside. It is often an area of mixed land use and rapid change

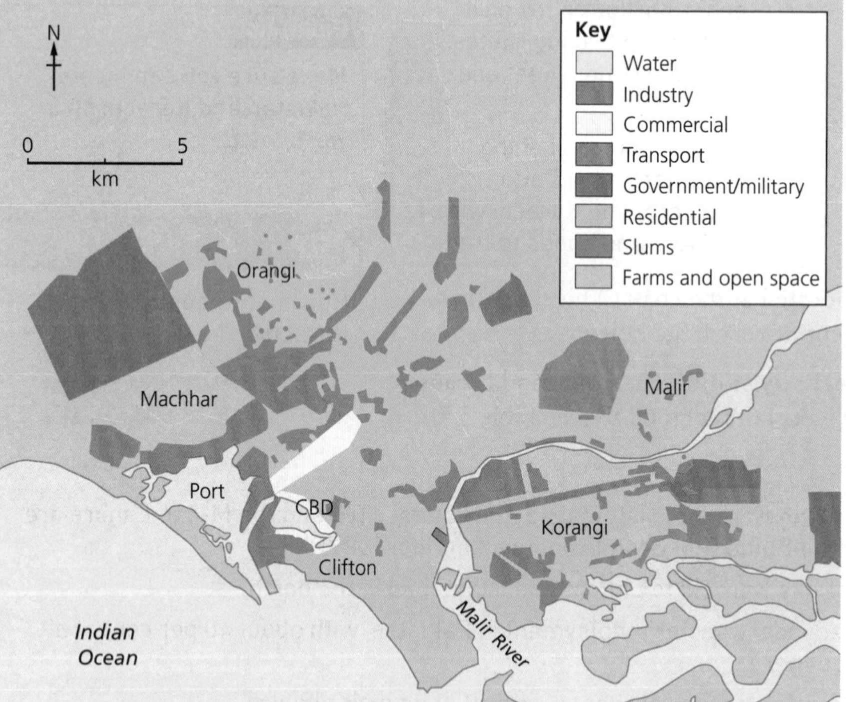

Figure 3.5 Karachi's urban structure.

Revision activity

Locate Karachi on Google Maps or Google Earth and zoom in to look at some of the areas in Figure 3.5.

Exam tip

You need to be able to name specific places within Karachi for the exam, because this city is a case study.

Exam practice

1 Define the term 'megacity'. [1]
2 Explain why both site and situation were important in the growth of a named megacity. [4]
3 Describe the pattern of land use shown in Figure 3.5. [4]
4 Assess the significance of a named megacity for its country and in a global context. [8]

ONLINE

Now test yourself

Where is most industry in Karachi located?

TESTED

Rapid growth

REVISED

Karachi has experienced very rapid population growth over recent decades, as is shown in Figure 3.6. The exact population of Karachi is not known because the last census of Pakistan was in 1998. Its current population size of 24.3 million in 2016 is an estimate.

Karachi has grown because:

- In 1947, when British India was divided into newly independent Pakistan and India, many Muslims living in what became India migrated to Karachi. These people are called Muhajirs and are a powerful economic and political force in Karachi.
- In the 1950s and 1960s, healthcare in Pakistan improved, lowering the death rate; the birth rate remained high so the country's population, and that of Karachi, began to grow very rapidly.
- Many Bangladeshis moved to Pakistan during and after the 1971 Indo-Pakistani war.
- Karachi has always been a magnet for rural–urban migration: poor people from rural areas are attracted to the megacity's opportunities and 40,000–50,000 new rural–urban migrants arrive every month.
- Karachi also attracts international migrants especially from Afghanistan (fleeing conflict), South Asia and Central Asia due to the large number of employment opportunities.

Future population growth of Karachi is likely. Pakistan's population is still growing by around two per cent a year due to a fertility rate of 2.7. This **natural increase** combines with rural–urban and international migration to give a growth rate of four to five per cent per year for Karachi.

Natural increase Population growth in a city caused by high fertility rates and children being born in the city

Export processing zones (EPZs) Industrial and business parks with low tax rates and few regulations, which are attractive to TNCs as locations for their businesses

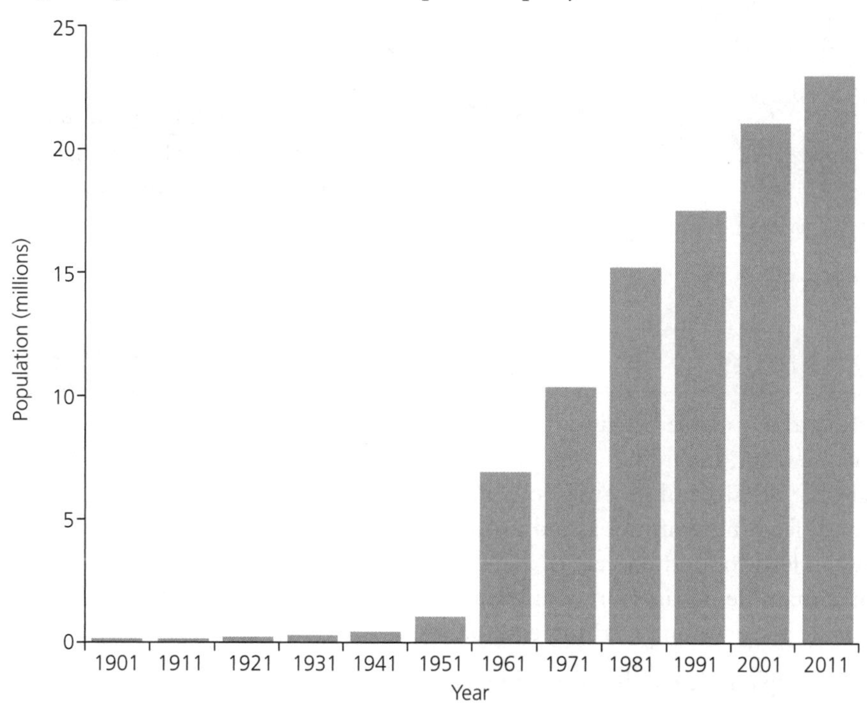

Figure 3.6 Population growth of Karachi from 1901 to 2011.

There are also economic reasons for Karachi's growth:

- Since 1980, Pakistan has used **export processing zones (EPZs)** in Karachi to attract foreign direct investment (FDI) from foreign TNCs, which has created jobs attracting migrants to the city.
- Pakistan's development of the CPEC is currently attracting large numbers of migrants building new infrastructure like power stations, motorways and railways.

Exam tip

Learn some facts and figures about Karachi's growth to add to your longer answers.

Now test yourself

How many rural migrants arrive in Karachi each month?

TESTED

Patterns and functions

Figure 3.7 shows how Karachi has grown from a small fishing and port town 80 years ago, into one of the world's largest cities:

- In the 1940s and 1950s, its spatial growth was simple: Karachi spread out from around the port area, with some **urban sprawl** along the north bank of the Malir River.
- The millions of people added to the population in the late 1950s and 1960s caused very large spatial growth north to New Karachi and east to Malir.

Urban sprawl The mostly uncontrolled growth of built-up urban areas outward from the city edge into the surrounding countryside

Pakistan industrialised in the 1970s and 1980s, fuelling greater rural–urban migration and population expansion:

- New industrial areas were developed east of the Malir River (Korangi Town, Landhi, Bin Qasim) including the Karachi EPZ area.
- Many city-edge slums, like Orangi Town, grew rapidly during this period.

Recent expansion has been to the north close to the M-9 motorway which is being built to link coastal Karachi with the cities of northern Pakistan.

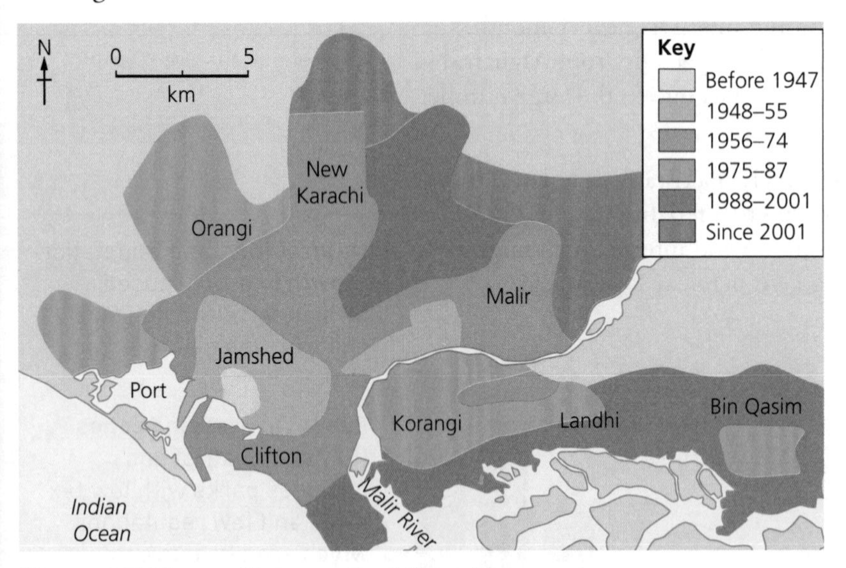

Figure 3.7 **The spatial growth of Karachi.**

Revision activity

Practise making a labelled rough sketch of Figure 3.7 that you could reproduce in the exam.

Opportunities and challenges

REVISED

Developing and emerging megacities attract people because they perceive the city as better than where they came from, including:

- Rural areas in the country, where life is hard and incomes low. People often struggle to farm enough food, farming is arduous, and drought, flood and other natural disasters can lead to food shortages.
- Other countries, which are poorer than the destination country and may have specific problems like war, persecution or failing economies.

These are all push factors: reasons for people to leave the place they come from. Megacities have many pull factors which draw people towards them:

- the chance of employment and a decent wage
- education resources like schools, colleges and universities
- healthcare resources like hospitals and clinics
- housing, leisure facilities and places to socialise.

For many people, especially low-skilled, poor, rural–urban migrants to developing world megacities, the fact that cities have these resources and opportunities does not mean they can access them.

Challenges

Karachi's very rapid population growth has created a number of challenges for people living in the city, especially for the low-income, recently arrived rural–urban migrants and their families. The problems are worse in the *katchi abadis* slums, like Orangi Town, that house 55 per cent of all Karachi's population (see Table 3.8).

Squatter settlements Slum housing built illegally on land owned by someone else

Table 3.8 The challenges facing Karachi

Housing shortages	There is not enough housing for all the new rural-urban migrants, which is why **squatter settlements** are built In the *katchi abadis*, eight or nine people live in each slum house on average Rents are very high for middle-class people because of the shortage
Squatter and slum housing	Orangi Town is Asia's largest slum, with 2.5 million people Settled illegally, it was never provided with roads and other services by the government Some areas have slowly developed into reasonable quality housing as residents have improved homes over the decades
Inadequate water supply	Only about 50 per cent of houses in Karachi's *katchi abadis* have their own piped water supply, and fifteen per cent are connected to a sewer Many people buy water from street sellers, costing $15–20 a month: much more than a piped supply at $5–10
Waste disposal	In the slums, the lack of a sewerage system leads to illnesses and disease Karachi has no city-wide system for disposing of its 8000 daily tonnes of garbage, so much of it is dumped in alleys and on waste ground In Korangi Town, 2500 factories and 200 leather tanneries dispose of industrial waste straight into the Indian Ocean
Limited services	75 per cent of *katchi abadi* houses have an electricity supply, but this is often an illegal, dangerous connection People have to pay for schools and health clinics, because there are not enough free government-run ones
Traffic congestion	Karachi has very bad traffic congestion, because most roads were designed for a much smaller city It has thousands of motorised rickshaws or *qingqi* that cause high levels of air pollution

Working life, especially for slum and *katchi abadi* dwellers, is very hard:

- Informal jobs such as street stalls, rubbish collection and recycling, transport and small workshops have long hours, low pay and very poor working conditions.
- Leather workshops (tanneries) and textile workshops making clothes are dangerous places to work, and child labour is common.
- Average household income in the slums is around $80–160 per month, so families struggle to survive.

Exam tip

Learn some data from Table 3.8. You need to have hard evidence of the poverty and challenges of living in Karachi.

Now test yourself TESTED

How many people are estimated to live in the Orangi Town slum?

Revision activity

Rank the challenges in Table 3.8 from most serious to least serious.

Wealth and poverty

Like all megacities, Karachi is a city of contrasts. It does have wealthy areas. Figure 3.8 shows that the areas with highest **quality of life** are located close the centre of Karachi and quality of life declines towards the most recently built slums and squatter settlements on the city edge.

Wealthy areas include:

- Clifton, on the coast south of the city centre. In Bahria Town, a recently developed **gated community**, villas and apartments sell for between $50,000 and $200,000.
- Karachi's formal workers live here. They work for banks, major TNCs like Toyota, and the government.
- The DHA (Defence Housing Authority) area is run by the Pakistani government and houses current and retired military personnel and their families.

There are mixed areas, like New Karachi. This is a suburb of formal houses and reasonably high incomes. These areas contrast with the slums on the city's edge like Orangi Town and Landhi.

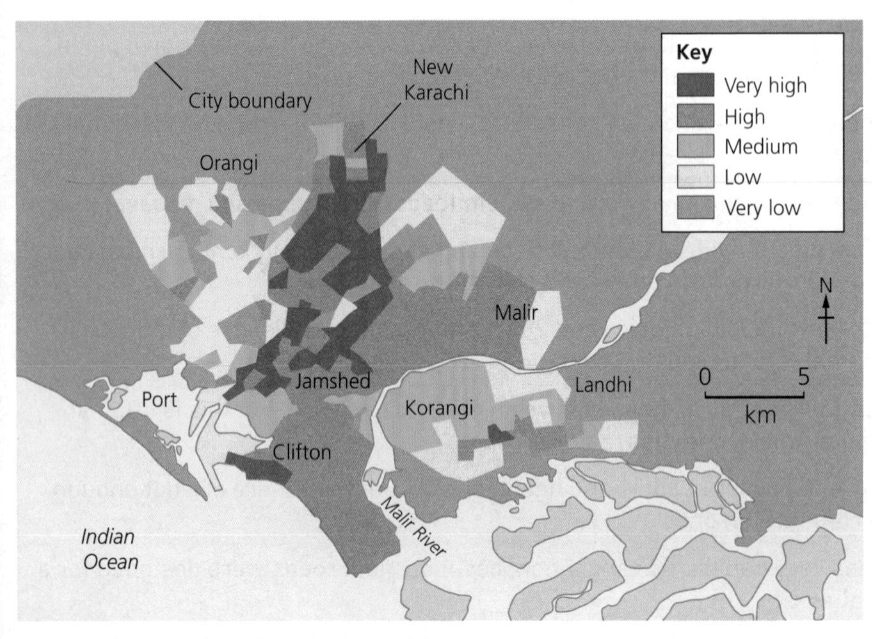

Figure 3.8 Quality of life in Karachi.

Managing a city like Karachi is a very difficult task:

- The rate of population growth, and slum growth, outstrips the ability of the city government of Karachi and its mayor to cope.
- Informally employed people pay no tax, so the city government income is low.
- Problems like lack of water supply, lack of sewerage and garbage are huge in a city of 24 million people.

Karachi's politics is also difficult:

- There were serious political riots in Karachi in 2007.
- The two main ethnic groups, the Pashtuns and the Muhajir, often disagree on policy.

Quality of life The housing, income, health and education situation of people, and whether or not these meet acceptable standards

Gated communities Private housing developments, with entry security and security patrols. Only residents can enter

Foreign aid Money either given or loaned by one country to another to help development

Now test yourself

What type of area is Clifton in Karachi?

TESTED

Exam practice

1 Describe the trends in Karachi's population growth shown in Figure 3.6. [2]
2 Suggest one economic reason for the rapid growth of a named megacity. [2]
3 Explain why the population of a named megacity has grown so rapidly. [4]
4 Describe the pattern of quality of life in a named megacity. [4]
5 Assess the challenges facing people living in a named megacity. [8]

ONLINE

Improving quality of life: top-down

REVISED

Karachi faces many challenges. A number of top-down projects funded by the city, the state government of Sindh, and **foreign aid** and loans aim to address these challenges (see Table 3.9).

Table 3.9 Top-down projects in Karachi

Project	Advantages	Disadvantages
Clean Drinking Water for All (2009), water supply	6000 water purification plants across Pakistan at a cost of £200 million Should improve formal supply and reduce disease The very poor may not be able to afford the water costs	Financed by US and Japanese aid, or loans for the World Bank and the Asian Development Bank Poorly planned, behind schedule, not providing water in new slums Total investment of only \$4 per person per year isn't enough
Waste collection by Wuzung, a Chinese company (2016)	Bins will be given to houses Free collection from households Should reduce disease risk	Currently only in south and east Karachi Bin lorries cannot access the poorest slums Any profit goes to Wuzung
Karachi Metrobus (2016), transport	Four rapid transport bus routes to carry 350,000 passengers per day Dedicated bus lanes will increase speeds and reduce congestion and pollution	Costs have doubled to \$244 million, and the project is delayed It is likely to benefit the middle class more than the poor

Karachi has serious air pollution. The United Nations' World Health Organization labelled it the fifth most polluted city in the world. However, a Japanese-funded air pollution monitoring system has not worked since 2012:

- Karachi Metrobus may help to reduce pollution be reducing the use of dirty motorised rickshaws.
- However, Karachi is part of the CPEC, which could bring even more people, transport and industry to the city – and more air pollution.
- CPEC is bringing \$33 billion of electricity supply investment, mostly funded by China. Some is renewable hydroelectric power, but most uses coal which will increase air pollution.

Now test yourself

TESTED

State two benefits of the Karachi Metrobus project.

Bottom-up

Top-down strategies are often not **sustainable strategies**, but some bottom-up ones may be. These are often led by community groups with help from non-governmental organisations (NGOs) and charities (see Table 3.10).

Table 3.10 Bottom-up projects in Karachi

Project	Advantages	Disadvantages
Orangi Town Pilot Project (since 1980), housing	Community group self-help project Helps groups of residents to build their own sewers in Orangi 90 per cent of Orangi's 8000 streets now have sewers The project creates community spirit, and reduces disease	Small budgets mean NGOs can only help small numbers of people Residents have to pay some of the cost themselves (about $50), and maintain the sewers Improved housing and sanitation doesn't mean improved incomes
Aman Foundation NGO (since 2008), health	Provides emergency ambulance services, maternal and other health services Focused on the poorest people in Karachi Provides a low-cost 'Telehealth' telephone service for the poor to get a diagnosis	Cannot meet all the needs of Karachi's poor As Karachi's main ambulance provider, it shows that the city government has failed to provide a good service Funded by donations and government grants, which can rise and fall
The Citizens Foundation (TCF), a Pakistani NGO (since 1995), education	Runs 150 schools in Karachi, for boys and girls Both primary and secondary education Scholarships are linked to income, so the very poor only pay five per cent of costs	NGOs rely on donations, which vary from year to year Doesn't help children forced to work to provide family income Despite its size, TCF has only 175,000 students

NGO and community-led projects are very important, but remember:

- Karachi is a city of 24 million people, and no NGO can help all of the millions who need help.
- Many NGO projects meet some basic needs (housing, water, basic education) but they don't necessarily increase wealth, which is needed if people's quality of life is to improve in the long term.

Sustainable strategies Ways of improving quality of life that will last a long time, are affordable, limit environmental problems and involve the community they are trying to help

Revision activity

Look at the top-down and bottom-up projects. Decide which are most and least sustainable in terms of helping poor people, and their environmental impacts.

Now test yourself

What percentage of Orangi's streets now have sewers?

TESTED

Exam practice

1 Define the term 'bottom-up strategy'. [2]
2 Explain two ways top-down strategies in megacities have helped low-income people. [4]
3 Assess how far bottom-up strategies to improve life in a named megacity have helped to improve quality of life. [8]

ONLINE

Exam tip

Make sure you can answer an exam question about assessing (weighing up) the advantages and disadvantages of top-down and bottom-up strategies.

Topic 4 The UK's evolving physical landscape

The physical landscape is a work in progress and constantly changing.

Why does the physical landscape of the UK vary from place to place?

A basic difference in the UK's physical landscape is between upland and lowland Britain. Broadly speaking, the boundary between the two is a line running from the Tees estuary in north-east England to the Exe estuary in Devon. To the north-west of this Tees–Exe line lies upland Britain, to the south-east lowland Britain (see Figure 4.1a).

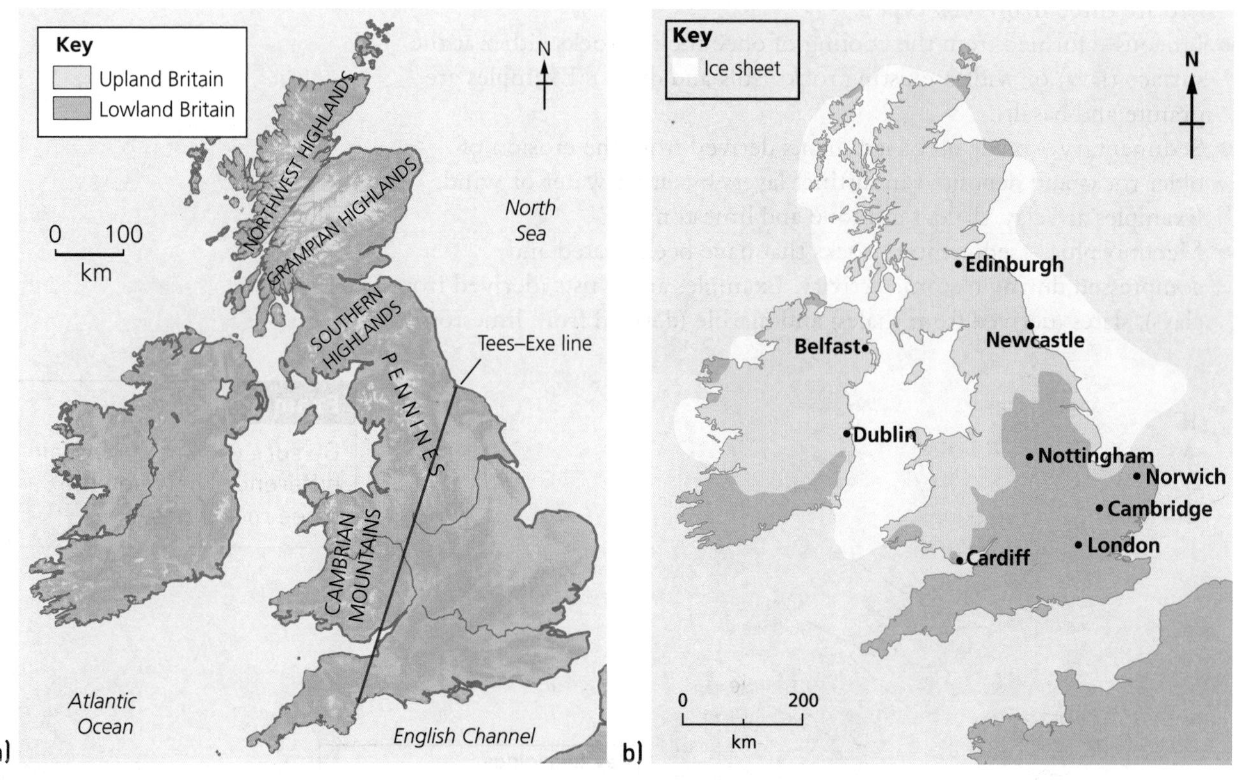

Figure 4.1 a) Upland and lowland Britain. b) Area covered by the last major ice advance over the British Isles.

Past processes and geology

REVISED

Processes

Throughout geological time, the climate of that part of the Earth's surface now occupied by the British Isles has changed. The evidence of much of this change lies in the rocks. But climates, past and present, also affect the processes at work on the physical landscape. The last and very spectacular changes in climate took place during the Pleistocene period. During this stage, there was a succession of glacial and interglacial periods. At the height of the glacial periods, much of the UK would have been covered by glaciers and ice sheets (Figure 4.1b). Those parts that were not covered experienced fluvioglacial and periglacial conditions. Again the Tees–Exe line is significant. It roughly marks the southerly limit of glaciation.

Tectonic processes have also contributed to the differences between upland and lowland Britain. Rocks in the former have been subject to much more folding and faulting.

Now test yourself

1 Name one location in upland Britain renowned for its glaciated landscape.
2 Name one landscape feature in lowland Britain that was produced by folding.

TESTED

Geology

There are three main rock types:

- Igneous – formed from the cooling of once molten rocks either at the surface (lava) or within existing rocks (sills and dykes). Examples are granite and basalt.
- Sedimentary – made up of sediments derived from the erosion of older rocks and deposited in distinct layers by either water or wind. Examples are clay, shale, sandstone and limestone.
- Metamorphic – sedimentary rocks that have been heated and compressed during tectonic activity. Examples are schists (derived from clays), slates (derived from shales) and marble (derived from limestone).

Exam tip

Be sure you know the main differences between the three rock types.

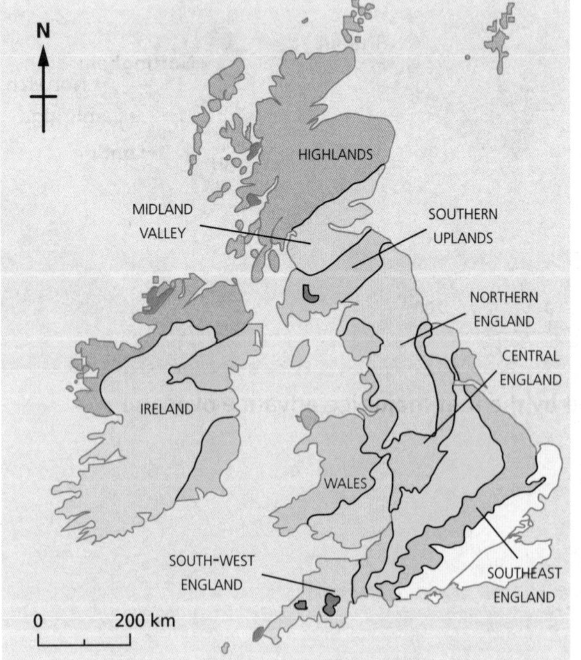

Key to geology
- Clays and sands
- Chalk
- Limestone, clay and shale
- Sandstones
- Limestones and sandstones
- Shales and limestones
- Igneous rocks and sandstone
- Granite (igneous) intrusions

Figure 4.2 Geological map of the British Isles.

Rock types are the key to understanding the differences in relief between the two parts of Britain. The higher relief and more rugged topography of upland Britain coincides with hard igneous and metamorphic rocks that are resistant to erosion. In contrast, much of lowland Britain is underlain by relatively soft and weak sedimentary rocks.

Table 4.1 shows geological and other important differences between Upland and Lowland Britain.

Table 4.1 Some generalised physical differences between upland and lowland Britain

Feature	Upland Britain	Lowland Britain
Altitudinal range	0–1400 m	0–250 m
Main rock type(s)	Igneous and metamorphic	Sedimentary
Geological age	Palaeozoic	Mesozoic and Cenozoic
Geological structure	Much folding and faulting	Some gentle folding
Processes during ice ages	Glacial	Fluvioglacial and periglacial
Climate	Wetter and more equable	Drier; greater temperature range

Exam practice

1 Name the two main factors responsible for the UK's uplands. [2]
2 Explain how landscape processes are affected by climate. [4]

Now test yourself

What were the three main processes that affected Britain during the Ice Age?

TESTED

Physical and human processes

REVISED

Upland and lowland landscapes

The physical landscape is the product of processes at work on different geological outcrops. But it is important to realise that those processes:

- have been at work literally for millions of years
- have varied over time with climate change.

The processes at work today are:

- weathering – breakdown of rocks in their original locations by the combined actions of the weather, plants and animals
- mass movement – the movement of weathered material down slopes under the influence of gravity
- erosion, transport and deposition by rivers.

Revision activity

You should have notes about two contrasting landscapes (one upland and the other lowland) which have resulted from these processes. Maybe you will find your two case studies in the following lists of popular choices:

- Upland – Dartmoor, the Lake District, the Yorkshire Dales, the Western Isles or the Cairngorms.
- Lowland – the Thames Basin, East Anglia, the Somerset Levels or the Weald.

Exam tip

Weather and mass movement are two important sub-aerial processes.

Case study: Dartmoor

Dartmoor is an example of an upland landscape formed when a massive dome of magma developed underground 290 million years ago. Joints were formed as the granite cooled and contracted. The presence of these joints made the rock vulnerable to freeze–thaw weathering, particularly during the Pleistocene ice ages. This eventually led to the break-up of the granite. Over time, weathering and erosion have exposed most of the granite. Mass movement processes have removed the broken granite downslope to form clitter slopes (Figure 4.3). The less-jointed and less weathered granite has been left upstanding as rounded peaks known as tors. The highest of these rises to 620 m above sea level. These continue to be the target of freeze–thaw and chemical weathering. Rivers have carved valleys radiating out from the centre of Dartmoor.

Figure 4.3 Dartmoor: a landscape of tors and clitter slopes.

Now test yourself

TESTED

1 What are the distinctive characteristics of granite?

Case study: the Weald

Strictly speaking, the Weald is an 'upland' landscape in lowland Britain. Its highest point is about 250 m above sea level. The Weald started life as a huge **anticline** (see Figure 4.4). The top of the anticline has been exposed to much weathering. The resulting regolith has then been eroded away by rivers. This has resulted in the exposure of strata of contrasting character. Resistant rocks, like the Chalk, form prominent escarpments. Flatter and lower vales have developed on the softer clays. The term used to describe this sort of landscape is scarp and vale topography. At the centre of the eroded anticline, the exposed sandstones give rise to an undulating landscape.

Anticline An up-fold of rocks resulting from compression in the Earth's crust

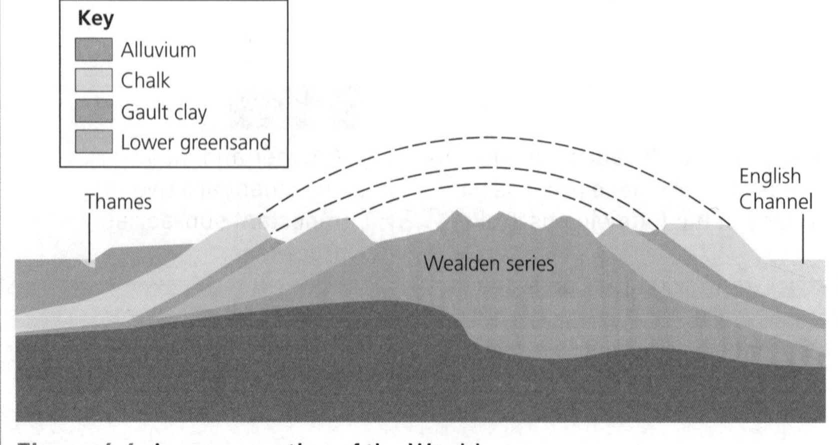

Figure 4.4 A cross-section of the Weald.

Now test yourself

2 Why is chalk so resistant to erosion?

TESTED

Case study: the Yorkshire Dales

The Yorkshire Dales are an upland area underlain by Carboniferous Limestone. Never densely populated, they are largely given over to livestock farming and more recently to tourism. The landscape of the Yorkshire Dales today is still largely determined by its geology and physical processes: the gorges, dales, waterfalls and limestone pavements. The building materials used in the old towns and villages, as well as in the barns and dry stone walls, have come from local quarries which pockmark the landscape. The bonus is that this use of local stone means that the settlements blend with the physical landscape.

Figure 4.5 A Yorkshire Dales landscape.

Case study: the Fens

The Fens are a flat and natural marshy area in eastern England. Formerly wooded, the **fens** were flooded by the post-glacial rise in sea level. Settlements at this time were located on small islands rising above the shallow waters. Local people made a living from fishing, trading by boat and digging peat. Land reclamation since the seventeenth century has produced a new and artificial landscape of large, fertile and flat fields. The Fens are now one of the most prosperous agricultural areas in the UK. Parts of the Fens lie below sea level and are protected from sea and river flooding by 250 km of embankment – a conspicuous landscape element together with the pumping stations and drainage channels.

Figure 4.6 An aerial view of the Fens landscape.

Impacts of human activity

In all the populated areas of the world, the landscape is not exclusively physical. There are biological elements (soils and ecosystems) and human elements contributing to the present landscape.

The human elements that make their mark on the landscape include:

- settlement – from farmsteads to cities
- economic activities – from agriculture and forestry to manufacturing and retailing
- transport – from roads and railways to ports and airports
- recreation and leisure – from public parks and playing fields to national parks and nature reserves.

Now have a quick look at two contrasting landscapes.

Fen A broad area of nutrient-rich shallow water.

Now test yourself

TESTED

Identify three landscape differences between the Yorkshire Dales and the Fens.

Exam practice

1 Assess the part played by geology in the creation of uplands in the UK. [8]

2 Explain two ways in which human activity has influenced the UK's physical landscape. [4]

ONLINE

Why is there a variety of distinctive coastal landscapes in the UK?

Physical interactions

REVISED

The coast is the boundary between the sea and the land. It is a particularly dynamic zone, always changing.

Coastal landforms

The most important factor affecting the coastal landscape is the geology:

- The resistance of different rock types to weathering and erosion. Hard rocks, such as granite and limestone, often give rise to headlands and wave-cut platforms. Soft rocks, on the other hand, are easily eroded, often becoming bays.
- The structure of rocks, particularly the arrangement of strata relative to the actual coastline. Where the rock structure runs parallel with the coast, a concordant coast of linear islands and sounds may be formed. Where the strata are at right angles, a discordant coast of alternating headlands and bays is the result.
- The presence of joints and faults in the rock. These are particularly vulnerable to erosion by the sea.

Now test yourself

TESTED

What is the difference between rock type and rock structure?

Case study: Isle of Purbeck coast

The Isle of Purbeck coast marks the eastern end of the famous Jurassic coast:

- Figure 4.7a shows alternating headlands and bays produced by alternating outcrops of hard and soft rocks. This is an example of a discordant coast.
- Figure 4.7b shows the part played by joints and faults in the evolution of Ballard Point with its sheer cliffs. Here caves give way to arches, and arches to stacks and stumps.
- Figure 4.7c shows the concordant south coast where the outcrop of limestone has been breached by the sea, and the weak clays behind it are being excavated into bays, the most famous of which is Lulworth Cove.

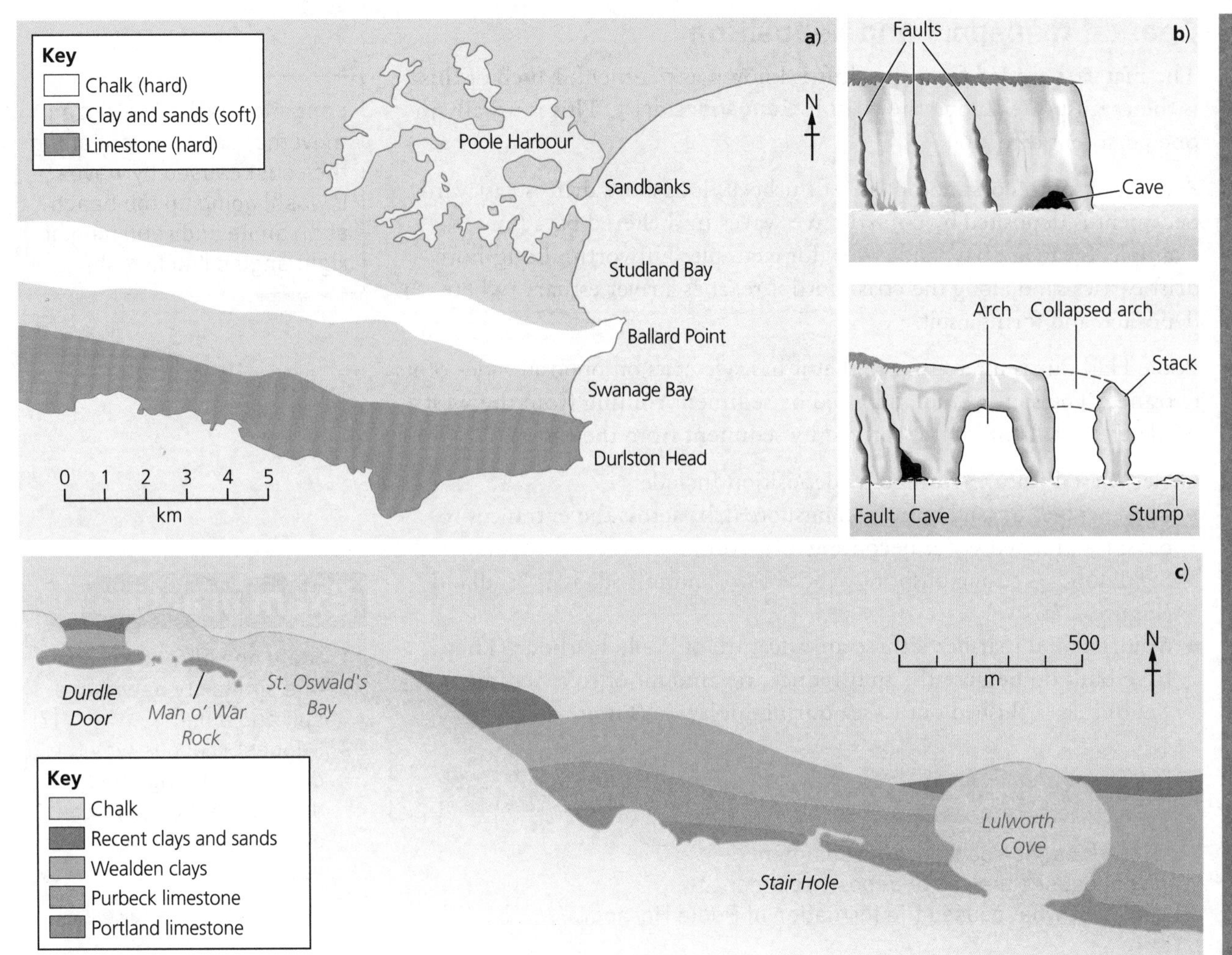

Figure 4.7 a) Isle of Purbeck; b) headland retreat at Ballard Point; c) bay development along the south coast.

Coastal erosion

The east and south coasts of the Isle of Purbeck are being eroded and are therefore retreating. Erosion is being encouraged by a number of factors including:

- The prevalence of destructive waves which enhance the processes of erosion, particularly abrasion and hydraulic action.
- The **sub-aerial processes** at work on the cliffs.
- Prevailing winds from the south-west mean that the south coast is very exposed; this is also the direction from which winter storms approach. It is not surprising that the limestone ridge has been breached.

Sub-aerial processes
A broad term that covers all the physical processes taking place on the Earth's surface, such as weathering, mass movement and erosion

Revision activity

Be sure you know the four erosion processes of the sea: abrasion, attrition, hydraulic action and corrosion.

Coastal transport and deposition

The material eroded from the cliffs is known as sediment. Much of this is then transported along the coast by **longshore drift**. This is usually in one persistent direction.

Along the south coast of the Isle of Purbeck, longshore drift is eastwards. Sediment is deposited by constructive waves in sheltered bays (for example, Swanage Bay) and coves (for example, Lulworth). Longshore drift carries sand along the coast until it reaches a river estuary, where it is deposited and forms a spit.

Poole Harbour is interesting in that it has two spits on opposite sides of its entrance. The spit at Studland is fed by sediment coming from the west, while the Sandbanks spit is formed by sediment from the east.

Other coastal features formed by deposition include:

- Bars – ridges of sand built by longshore drift across the entrances to bays; the trapped water becoming a lagoon.
- Sand dunes – strong onshore winds blow sand inland, as at Studland (Figure 4.7a).
- Mudflats and marshes – these are a feature of Poole Harbour. They have built up because the spits reduce the amount of river sediment (mud) that is flushed out to sea during high tide (Figure 4.7a).

Longshore drift The zigzag movement of sediment along the coast caused by waves (swash) going up the beach at an angle and returning at right angles (backwash)

Now test yourself

TESTED

1. How does the sea transport sediment?
2. What causes the sea to deposit sediment?
3. Suggest what caused the formation of Poole Harbour.

Exam practice

1. State one difference and one similarity between a spit and a bar. [2]
2. Suggest reasons why the direction of longshore drift along the south coast of the Isle of Purbeck is eastwards. [4]

ONLINE

Human interactions

REVISED

Around the coast of the UK, there are few stretches that have been left untouched by human activity.

Activities and their impacts

As far as the coast is concerned, human activities fall into two categories (Table 4.2):

- those that are land based
- those that take place offshore.

Table 4.2 Some coastal activities

Land-based activities	Offshore activities
Agriculture	Commercial fishing
Energy and industry	Dredging (sand and gravel)
Housing	Recreation (boating, sailing, fishing)
Offices and services	Shipping
Tourism	Tourism
Transport	Wind farms

Clearly, all the land-based activities alter the visual appearance of the coastal landscape in some way or another. The most obvious are those activities that involve large structures, such as power stations, wind farms, oil refineries and high-rise apartment blocks.

There are other activities that can literally change the physical configuration of the coastline, such as port and industrial developments (see the Southampton Water case study on page 72) and coastal management.

An obvious impact of any new activities, such as industry, offices and tourism, is to increase the number of people in the coastal zone. This in turn means building more houses, schools, medical centres, better roads, and so on. All of these will change the coastal landscape.

All of these land-based activities are competing for space along the coast. There is also potential conflict between some of them, as for example between industry and tourism.

With the exception of wind farms, the visual impact of offshore activities is minimal. But, as on land, there is both competition and conflict.

Now test yourself

TESTED

1. Which activities compete for seafront locations?
2. Identify two conflicting coastal land uses.
3. Why might there be objections to the building of an offshore wind farm?

Coastal change

Most coastal change falls into one of two categories:

- erosion and retreat – there many examples in UK (see Figure 4.9, page 73)
- deposition and advance – there are few examples.

In addition to these basic physical changes, coasts are being modified by those activities listed above.

Exam practice

1. Identify the land-based activity that you think most affects the coastal landscape. Explain your reasons. [4]
2. Assess the arguments for and against making use of coastal mudflats and marshes. [8]

ONLINE

Exam tip

You need to have studied a coastal landscape that has changed, and the physical and human processes causing that change.

Case study: the changing face of Southampton Water

Located at the head of Southampton Water, Southampton is one of the UK's leading ports. Originally, the shores were fringed by mudflats and salt marshes.

In the nineteenth and twentieth centuries, mudflats were reclaimed on which to build new docks for the large ocean-going liners carrying passengers to all parts of the world. These docks were later extended upstream to provide the large terminal facilities needed for handling containerised cargo. Southampton is now the UK's second busiest container port.

In the middle of the twentieth century, more land was reclaimed on the southern shore close to the entrance to Southampton Water to provide the site for a large oil refinery. Later, even more land was reclaimed for an oil-fired electric power station. At the same time, salt marshes were reclaimed at Dibden Bay to accommodate a planned further expansion of the port. A small area at the southern end of Dibden Bay has been used as the site for a residential marina at Hythe. Hamble, on the other side of the water, is a renowned yachting centre.

Most of the salt marshes have now been reclaimed and other sections of the Southampton Water shores have become changed by suburban settlements such as Marchwood, Dibden Purlieu and Fawley. In fact, all this development of the Southampton Water coastline has led to a range of different land uses being accommodated.

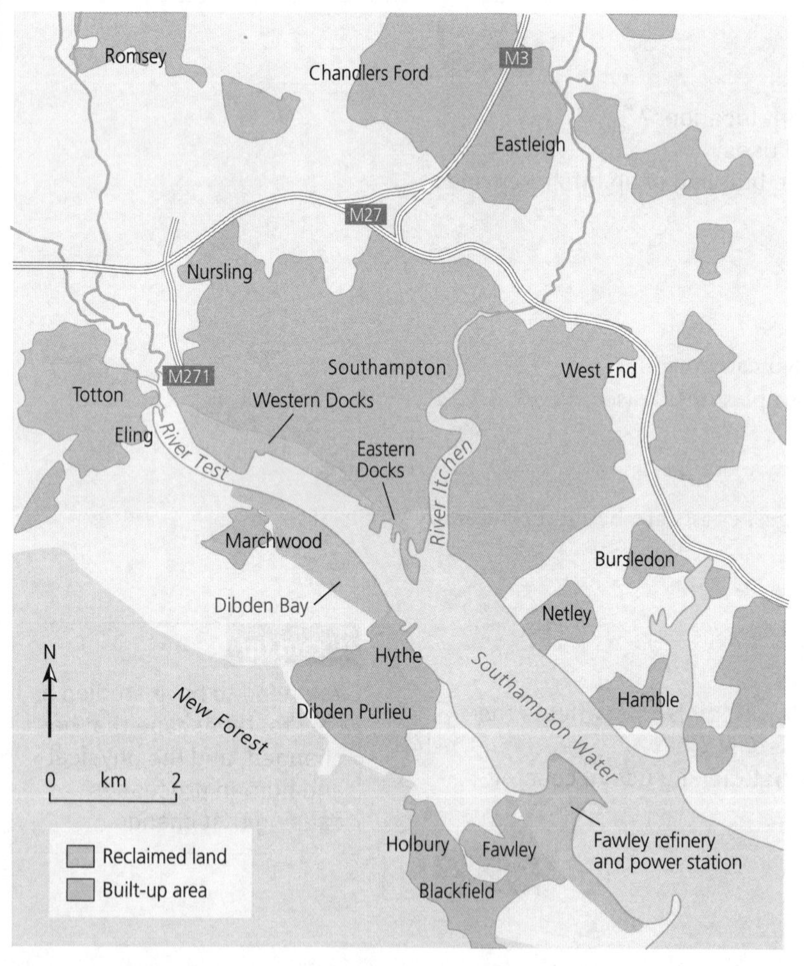

Figure 4.8 Southampton Water.

Unfortunately, the natural environment and conservation have taken a back seat. It is only recently that parts of the western shore have received some official protection. But this protection has come far too late: it was needed 100 years ago. One particular environmental challenge at the moment is what to do with the now closed Fawley power station which dominates the coastal landscape at the entrance to Southampton Water.

Now test yourself

TESTED

Why was the original shoreline of Southampton Water mainly mudflats and salt marsh?

What are the challenges facing coastal landscapes and communities?

Coastal challenges and management options

REVISED

Increased coastal flooding

A major concern in some coastal areas of the UK is the risk of increased erosion and flooding (see Figure 4.9). The risk is related to global warming, but in two different ways:

- the rise in sea level which threatens to drown coastal areas
- the increased frequency of damaging storms and storm surges which are capable of breaching coastal defences.

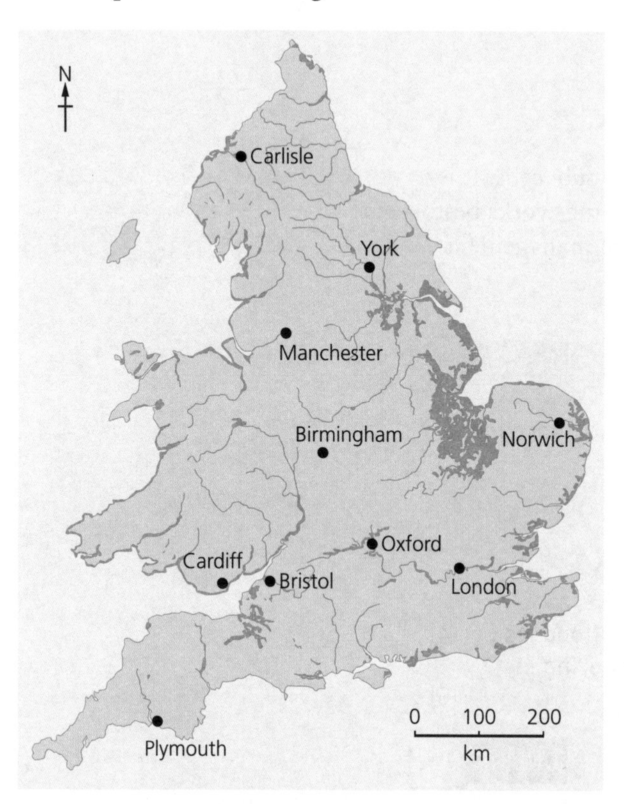

Figure 4.9 Stretches of coastline in England and Wales most at risk from erosion and flooding.

Revision activity

Study Figure 4.9 and make notes about those parts of the coast most at risk from erosion and flooding.

Just remind yourself of how important the UK's coastal areas are:

- Many of our cities have a coastal location.
- A significant percentage of the population lives within sight of the coast.
- Coastal lowlands provide valuable agricultural land.
- The coast has attracted a fair amount of industrial development.

Managing coastal processes

How should the threats posed by global warming to this important UK fringe be managed? Coastal managers have three options:

- Hold the line – build up sea defences to prevent coastal erosion and flooding. So the coastline stays where it is. A great deal of money will be needed to constantly upgrade the defences.
- Managed retreat (also known as strategic realignment) – gradually let the coast erode, but move businesses and people away from those areas most at risk. Less money is spent on sea defences, but money is needed to compensate those who have to move.
- Do nothing – let nature takes its course and be prepared to suffer losses. This could be expensive in terms of paying out compensation.

Today, the UK's approach to coastal management has become much more comprehensive and holistic. It is known as integrated coastal zone management (ICZM). It brings together all those with an interest in the coast (stakeholders) to plan a way forward that:

- resolves conflicts
- works with natural processes
- respects the natural environment and its ecosystems
- is transparent and sustainable.

The UK's coastline is divided up into cells, and these into sub-cells. It is at this level that integrated coastal management and planning works best. Every sub-cell is required to prepare and agree a shoreline management plan.

Exam tip

Be sure you can name a stretch of coastline in the UK where there is managed retreat.

Now test yourself

TESTED

What are the distinguishing features of a soft engineering approach to coastal management?

Exam practice

1 Explain why global warming is increasing the coastal flood risk. [4]
2 Assess the costs and benefits of hard and soft engineering in managing coastal erosion in the UK. [8]

ONLINE

Case study: holding the line in Christchurch Bay

Christchurch Bay lies to the west of the town of Christchurch on the south coast of England (see Figure 4.10). It is bounded on the west by Hengistbury Head and by Hurst Castle Spit on the east. It is a sub-cell within the cell stretching from Portland Bill (Dorset) to Selsey Bill (West Sussex).

Christchurch Bay has a retreating coastline, but the fact that is also a very urbanised coast makes holding the line the necessary management strategy. The cliff retreat hotspot is Barton-on-Sea. Figure 4.11 shows the main reasons for this.

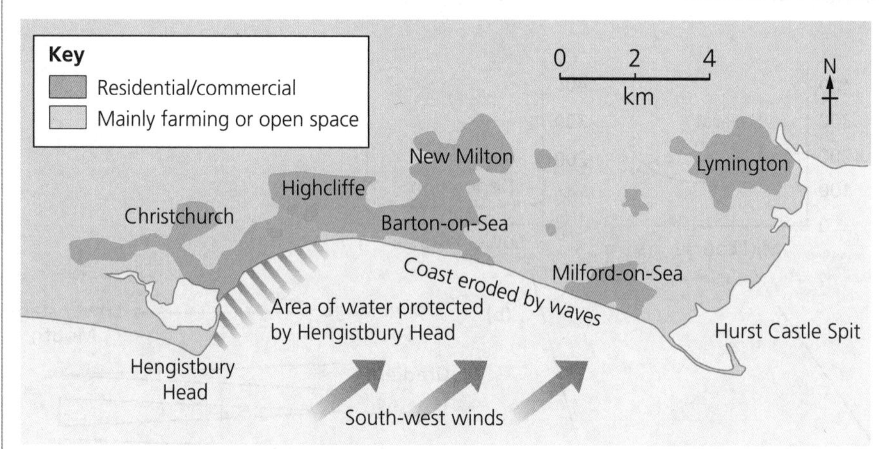

Figure 4.10 Christchurch Bay.

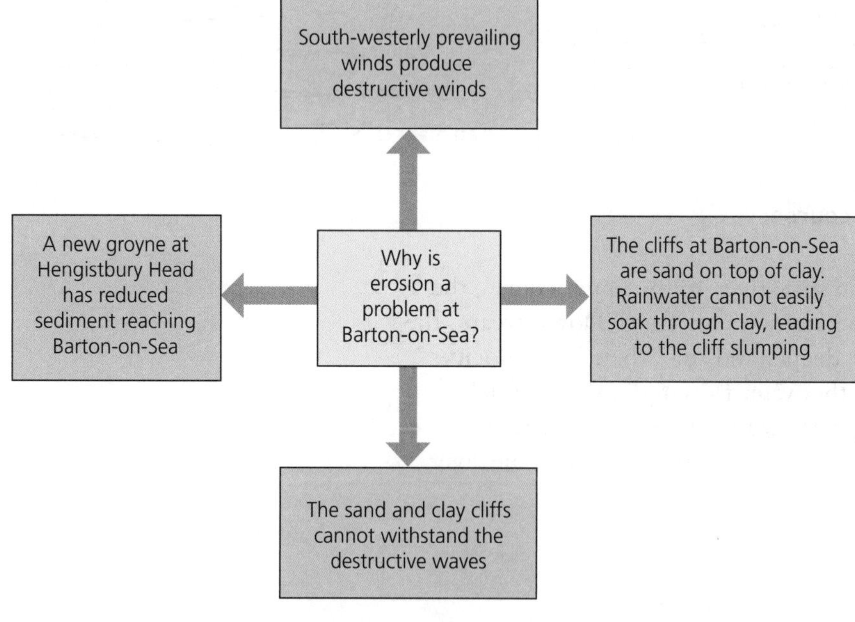

Figure 4.11 Causes of cliff retreat at Barton-on-Sea.

Christchurch Bay is a popular residential and retirement area attracting people from all parts of the UK. With so much housing development close to the cliff edge, there has been no option but to use hard-engineering techniques to hold the line:

- Bull-nose concrete sea walls are used to deflect waves back out to sea.
- Rip-rap is used to create offshore breakwaters that weaken the waves before they hit the sea wall.
- Groynes have been built to retain beaches (a natural sea defence) as much as possible. The problem is that while the groynes trap sediment in one place, they stop it reaching another in the direction of the longshore drift. One impact of this is that Hurst Castle spit is being deprived of the shingle needed for its survival.

The only soft engineering methods used here are:

- Beach replenishment – sand and shingle are pumped on to the beaches to increase their size.
- In-cliff drainage – drainage pipes are drilled into the cliffs so that the cliffs do not become saturated and prone to slumping. This is also known as slope stabilisation.

Maintaining the coastal defences around Christchurch is expensive, but the costs of holding the line are justified by the huge sums of money invested in the cliff-top urban development. In this instance, all stakeholders benefit.

Now test yourself

TESTED

What are the benefits of integrated coastal zone management?

Why is there a variety of river landscapes in the UK?

Different characteristics

REVISED

You should have studied the course of one named river to understand how and why river landscapes change between its upper and lower courses.

Downstream changes

The long profile of a river runs from its source to the point where it enters the sea or a lake or joins another and larger river. The character of the river channel and its valley changes downstream (see Figure 4.12).

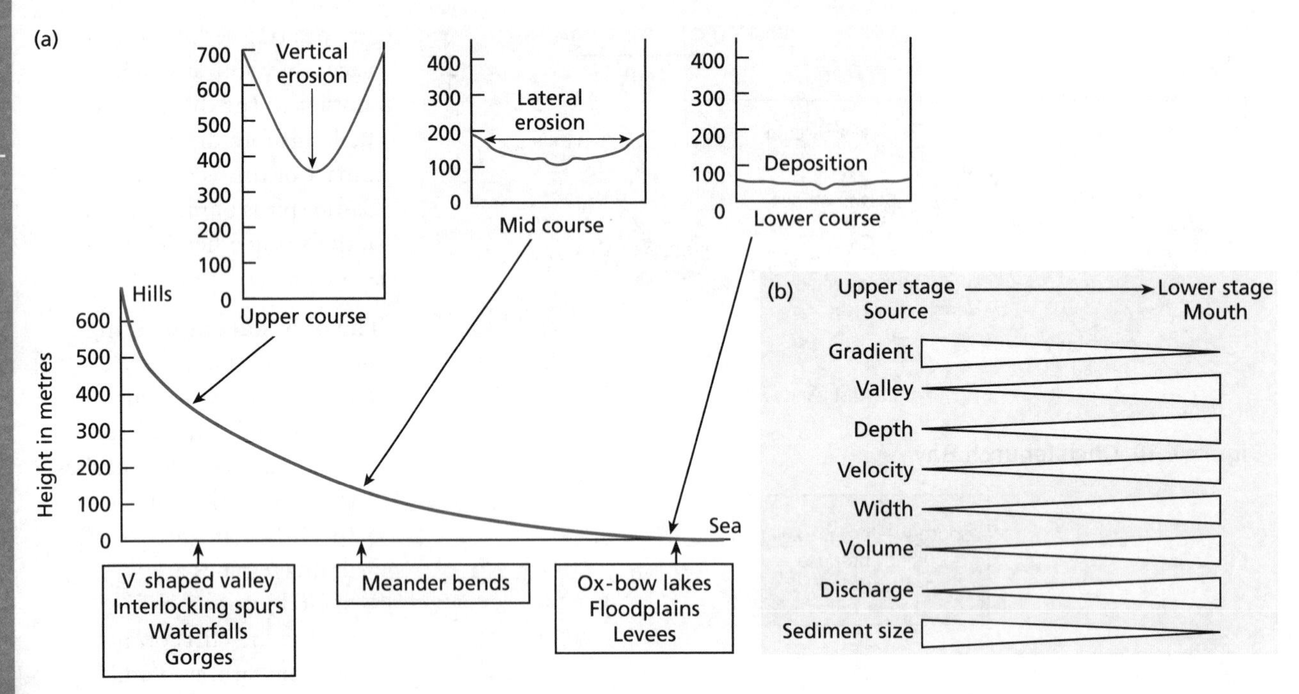

Figure 4.12 Features of a river's course.

The long profile comes gentler and smoother. In the upper course, the river channel takes up most of the narrow valley floor. Downstream, the channel gradually becomes wider, deeper and smoother. The velocity (speed) and discharge (volume) of the water flowing in the channel continue to increase despite the gentler gradient. Sediment load increases downstream, but the sediment size diminishes. In plain view, the river channel becomes less straight.

Now test yourself

TESTED

Why does sediment size decrease downstream?

River processes and landforms

River processes are very similar to those of the sea, namely:

- erosion – hydraulic action, abrasion, attrition and solution
- transport – traction, saltation, suspension and solution
- deposition – this occurs when there is a decrease in the energy, speed and discharge of the river.

The river processes, along with weathering and mass movement, produce distinctive landscape features in the three sections of the long profile (see Figure 4.13).

Upper section

- Steep-sided V-shaped valleys – resulting from vertical river erosion plus weathering and mass movement on the valley sides.
- Interlocking spurs – projections of higher ground that alternate from either side of a valley and point towards the valley floor.
- Waterfalls and rapids – formed where rivers flow across bands of hard rock that are more resistant to erosion.

Middle section

- Meanders – sinuous river channels resulting from bank erosion at alternating points along the channels.

Lower section

- Floodplains – valley floors over which rivers spread during floods.
- Levees – ridges of sediment built up along the banks of river channels.
- Oxbow lakes – crescent-shaped lakes or marshy areas occupying former meanders that have become cut off by the breaching of meander necks.

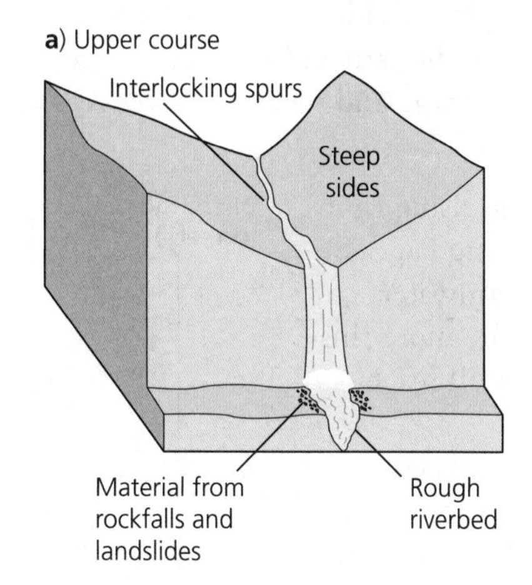

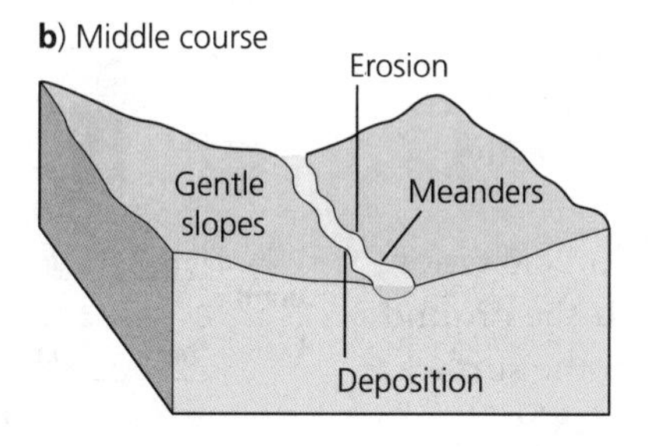

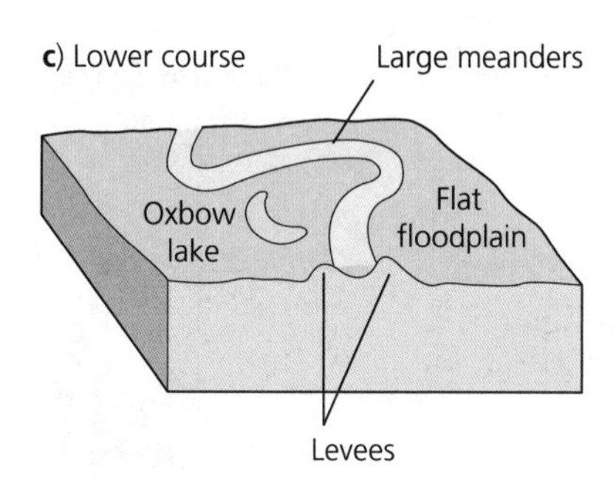

Figure 4.13 Some river landforms.

If the river ends at the sea or a lake, the floodplain spreads out to become either a delta or an estuary.

Hydrographs

A hydrograph shows how a river's discharge varies over time. Rises in discharge are related to periods of rainfall or possibly to the melding of snow and ice. A critical feature of a hydrograph is the time lag between peak rainfall (the highest amount of rain falling in a unit of time) and peak discharge (the highest rate of discharge following a rainfall event) (see Figure 4.14). The hydrograph curve divides into two parts:

- a rising limb – between the start of the rain and peak discharge
- a falling limb – between peak discharge and when the river returns to its original discharge.

Revision activity

Be sure you have some brief notes about how all these landscape features are formed.

The time lag and the gradient of the rising limb are affected by a number of interacting factors:

- The types of flow carrying the rainwater from where it fell to the river – the more this is by surface runoff, the faster it will reach the river.
- The intensity of rainfall – heavy rain will not sink into the ground; instead it will become overland flow and quickly reach the river.
- The antecedent conditions – what the ground conditions were like before the particular storm: was the ground already saturated or was it very dry?
- Temperatures affect the form of precipitation – snow can take weeks to melt. If the ground remains frozen, melting snow on the surface can eventually reach the river quickly.
- Geology – hard, impermeable rocks will mean more surface runoff; with porous rocks water will move more slowly via throughflow and groundwater flow.
- Slopes – steep slopes mean more runoff and a fast, direct delivery of water to the river.
- Drainage basin shape – in an elongated drainage basin, the time lag will be less than in a circular one.
- Soil type – its degree of permeability; how quickly rainwater passes down through the soil.
- Land use – a dense vegetation cover will intercept and delay the rain reaching the ground. However, the tarmac and paved surfaces, building materials and drains of urban areas greatly speed up runoff (see below).

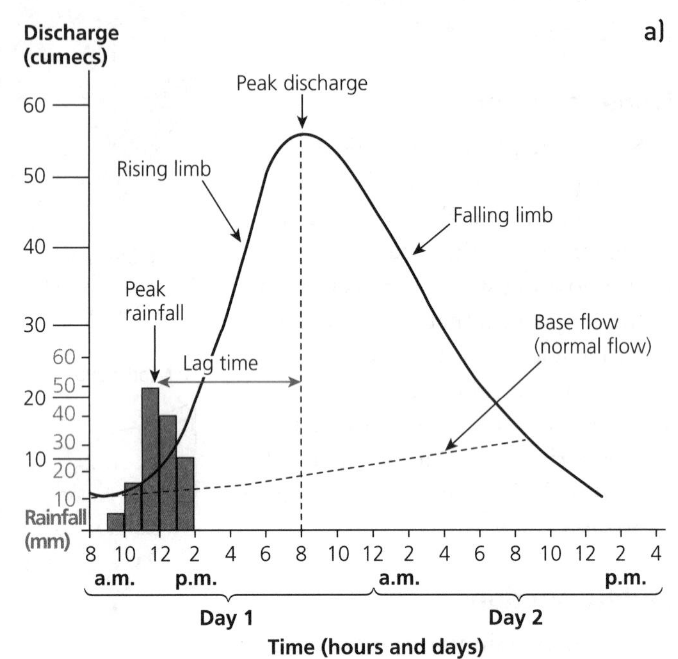

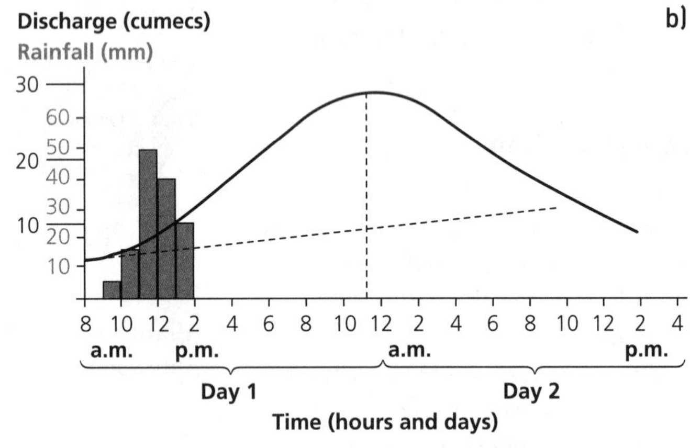

Figure 4.14 Two hydrographs: a) fast response and b) slow response.

Hydrographs may be classified as having either a flashy (fast) or a slow response to a period of rain (see Figure 4.14).

Now test yourself

TESTED

1. What are the four important components of a hydrograph?
2. How does drainage basin shape affect time lag?
3. Why is the rising limb often steeper than the falling limb?

Exam practice

1. Explain the differences between an estuary and a delta. [4]
2. Assess the value of hydrographs. [8]

ONLINE

Impact of human activities

REVISED

Impacts of river landscapes and hydrographs

This topic can be illustrated by focusing on two land-use changes:

- Urbanisation. This greatly alters the nature of the ground. During heavy rainfall, impermeable surfaces such as roofs, pavements and roads encourage surface runoff and allow little infiltration. Drains also speed up the delivery of rainwater to the river. In short, as a result of urbanisation, the hydrograph will become a quick response one (see Figure 4.14a). Urbanisation is also likely to change river landscapes in many ways, but particularly near to any rivers that have a flood risk.
- Deforestation. With trees removed, there will be less interception and more rainfall reaching the ground. Despite increased infiltration, more of the rainfall will reach the river and possibly more quickly than before. The rising limb will become steeper. There is also likely to be soil erosion, particularly on valley sides. The eroded soil will add to the river's load and possibly lead to some deposition. In short, the slow-response hydrograph typical of a forested area (see Figure 4.14b) will change and show a steepening of the rising limb.

Now test yourself

TESTED

What do these two land-use changes have in common in terms of changing hydrographs?

Exam tip: Investigating UK Geographical Issues questions

In Paper 2 there will be two UK Geographical Issues questions worth 8 marks: question 4 and question 7. These questions have a synoptic aspect to them, because they link different areas of geography. One of these questions will have an extra 4 marks from spelling, punctuation, grammar and specialist terminology. Both questions will:

- use the command word 'assess' – so you will need to weigh-up different factors, make judgments and come to a conclusion
- use a figure, like a map, table of data or graph
- you will need to look closely at this figure and analyse it – pick it apart and look for trends, patterns, anomalies – perhaps even do some simple calculations like working out a mean
- each question will link together two areas of geography (this linking synoptic); for instance, a question could combine the physical factors that increase flood risk (geology, flat land, floodplains) with human factors (high population density, impermeable surfaces)
- your answer will need to refer to the figure, both areas of geography and any other information you think is relevant; for instance, other examples you know of (this is synoptic).

River flooding

Case study: the Somerset Levels

This is a story of not one, but a number of rivers. They help to drain this huge lowland basin lying between the Mendip and Quantock Hills. The Somerset Levels are claimed to be one of the lowest areas in the UK. Much of the area lies below the high-water level of spring tides. All the rivers flow into the Bristol Channel.

In January 2014 the Somerset Levels experienced flooding of an extent and depth unknown in living memory. It is estimated that at one time ten per cent of the area was under water.

Physical causes

A quick but prolonged succession of Atlantic storms with gale-force winds and persistently heavy rain were the major cause. So much rain fell that the rivers could not cope. Particularly high tides in the Bristol Channel and its narrowing created great tidal surges. These effectively ponded back the floodwater trying to escape the levels. The good news was that the coastal defences coped with the tidal surges.

Human causes

Over the past few years before the 2014 floods, there had been less dredging of the rivers channels and drainage ditches than usual. Without dredging, the beds of river channels are gradually raised by the gradual accumulation of silt. This is turn means that a river more easily overflows its levees.

In the case of the January 2014 floods, it was not possible to identify any other contributory human actions – no major land-uses changes, no significant urbanisation or deforestation.

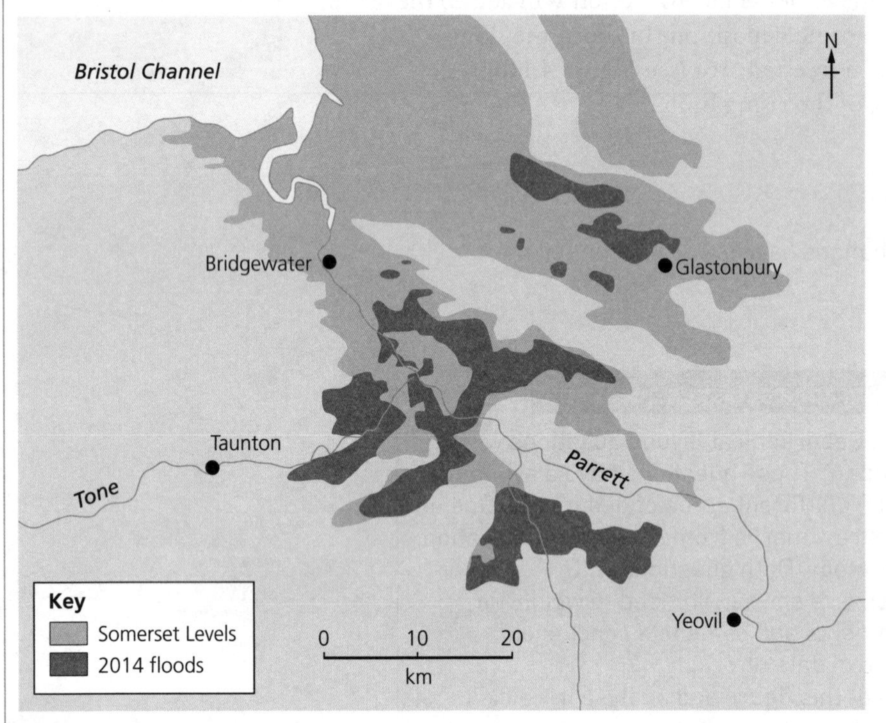

Figure 4.15 The Somerset Levels and the areas underwater in January 2014.

Now test yourself

TESTED

1. What is most noticeable about the distribution of flooding shown in Figure 4.15?
2. Do floods always have to have human causes?

Exam practice

1. Identify **two** river landscape changes resulting from deforestation. [2]
2. Assess the ways in which people can increase the risk of flooding. [8]

ONLINE

Exam tip

Be sure you have notes about flooding on one named river. It is important to distinguish between physical and human causes.

What are the challenges facing river landscapes and communities?

Some rivers are more prone to flood than others

REVISED

It is a fact of life that some rivers are more prone to flooding than others. The basic reasons for this are:

- genetic – to do with their basic physical characteristics and those of the basins they drain
- climatic – to do with global warming
- human – changes in land use that alter runoff and therefore storm hydrographs.

Increased river flooding

The causes of an increase in the risk of flooding are basically climatic (global warming leading to rising sea levels and increased storminess) and human actions and non-actions (see the Somerset Levels case study above). An increase in the risk of flooding calls for making some decisions about flood management. There are three broad management strategies:

- Do nothing – cheap but likely to mean a flood risk.
- Continue with present flood defences – they may be adequate now, but will they be in the future?
- Adopt a new flood management scheme – needs to be fit for purpose, that is able to cope with the heightened flood risks that are forecast. Certainly an expensive option.

Managing the flood risk

As at the coast, managing the river flood risk has both 'hard' and 'soft' solutions.

Hard engineering

Solutions include building:

- flood walls along river banks or raising the height of natural levees
- flood relief channels – creating extra channels to divert excess water away from high-risk areas
- flood barriers – upstream storage dams that hold back some of the floodwater so that it does not overflow river banks further downstream; also barriers near the river mouth to prevent a storm surge or high spring tide from flooding the river upstream of the barrier
- demountable flood barriers – put up when a flood is forecast, then taken down afterwards; because of this, they are less visually obtrusive.

These solutions are all very costly and therefore tend only to be used to protect valuable areas, such as city centres, heritage sites and populated high-rise areas.

Now test yourself

Give two reasons why people will be at greater risk of flooding in the future.

TESTED

Soft engineering

Solutions include:

- floodplain retention – clearing most forms of development away from floodplains and letting floodplains perform their natural function of temporarily holding floodwater until the river returns to normal flow
- floodplain zoning – using the floodplain in a way to minimise the impacts of flooding (see Figure 4.16)
- river channel restoration – returning rivers to their original state by taking away embankments and restoring meanders; this allows rivers to flood but slows them down
- dredging – making sure that the river channel is clear and so maximising river discharge; perhaps lining the channel with concrete.

Common to both hard and soft options, a good flood warning system must be a top priority.

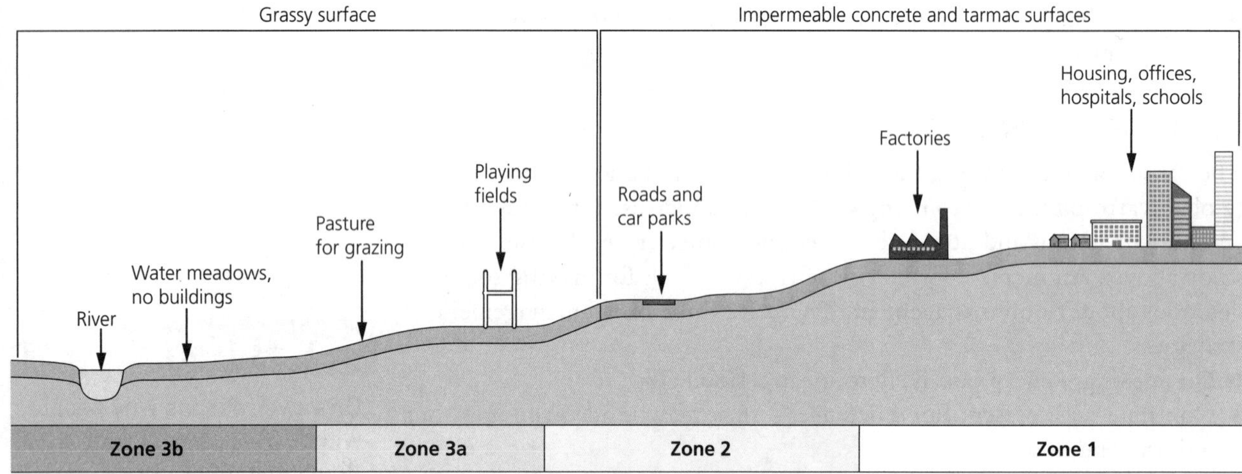

Figure 4.16 Floodplain zoning.

Now test yourself

TESTED

Suggest some ways people might improve the flood-proofing of their homes.

Overall, soft engineering solutions have the benefit of being cheaper and having less impact on river landscapes. However, they are thought to be less effective than hard engineering solutions when it comes to giving protection to high-risk areas during times of exceptional flooding. But the reality of the situation that everyone must face is this: there will be floods of such magnitude (say those once in a 100-year events) that even the most expensive and sophisticated river walls and flood relief channels cannot cope with.

Exam practice

1 Suggest reasons why some rivers are more prone to flood than others. [4]

2 Explain why soft engineering is often preferred to hard engineering when managing river flood risk. [4]

ONLINE

Topic 5 The UK's evolving human landscape

Why are people and places changing?

The human landscape of the UK has been changing for thousands of years. For most of that time the process of change has been a slow one. Over the past 250 years, however, the pace of change has accelerated.

Key elements of the human landscape

REVISED

Three are three main elements in this changing landscape:

- population – the growth in numbers and the changes in distribution
- economic activities – the shift in what people do for a living and what drives the country's economic development
- settlements – that house this growing number of people and many of the country's economic activities.

Urban cores to rural peripheries

In terms of the degree to which people have changed the physical landscape, there is a gradient, often referred to as the **rural–urban continuum**. It runs from remote areas that have been changed little by people to the centres of the **urban cores** focused on the UK's largest cities, such as London, Birmingham, Manchester and Glasgow, and their immediate **spheres of influence**.

Rural–urban continuum The unbroken transition from remote rural areas to the centre of an urban core

Urban core A densely populated area of the UK focused around one of the major cities

Sphere of influence The area which is dominated by the economic, social and political pull of an urban core

Table 5.1 summarises some of the basic differences between urban cores and rural areas.

Table 5.1 Comparison of urban cores and rural areas

	Urban cores	Rural areas
Population	Density of over 1000 persons per km^2	Density less than 200 persons per km^2
	Mainly young adults, children and single people	Mainly older people; few children
Economic activities	Mainly shops, offices and factories	Mainly agriculture; some tourism
Settlements	Conurbations and large cities	Market towns, villages and farms
	Some high-rise buildings	Low-rise buildings
	Property relatively expensive	Property generally cheaper
Examples	Greater London; Southampton	Bedford; Sharnbrook

It is the view of many that the UK's urban cores are too attractive and powerful. They dominate all aspects of the UK – the economy, population and government. Today, there are both government (for example, enterprise zones) and European Union (EU) schemes (for example, the European Regional Development Fund) aimed at reducing the gap between the urban cores and remoter rural areas, between the rich and the poor parts of the UK.

Exam practice

1 Suggest two reasons why population density varies within the UK. [4]
2 Explain how investment in transport might reduce the gap between urban cores and rural areas. [4]

ONLINE

Now test yourself

1 What is the difference between a conurbation and an urban core?

TESTED

Links to the wider world

REVISED

The UK links to the rest of the world in a number of different ways. Two of those ways are in focus here, namely population migration and economic production. Globalisation and the growth of the global economy are increasing the interdependence of countries, particularly through trade and investment.

Migration

The UK covers an area of only 244,000 km^2 but its population has grown by 10 million over the past 50 years (by nearly twenty per cent). That growth was the outcome of two processes:

- natural increase – where births exceed deaths
- net immigration – where the number people entering the UK is greater than the number leaving.

The birth rate has begun to rise over the past ten years. This and a fairly stable death rate mean that almost half of the UK's increase in population is now due to natural increase.

The present volume of immigration into the UK has become an issue. The UK's immigrants come from two main sources:

- Commonwealth countries (former British colonies) mainly in the Indian subcontinent, Africa and the Caribbean – these migration flows have been encouraged by the UK government
- EU countries – being in the EU allows a country's citizens to move freely between member countries. Significant here has been the influx of workers from east European countries. This may well change after Brexit (when the UK leaves the EU).

With both sources, the major pull factor has been employment. A significant outcome of this net immigration has been an increased ethnic and cultural diversity. This is especially true of those coming from Commonwealth countries. Even so, these groups only account for a little over ten per cent of the UK's population. Furthermore, these groups are not evenly distributed throughout the UK. They are concentrated in the urban cores, and in particular areas within those cores. In short, there is a segregation of different ethnic and cultural groups.

Now test yourself

2 What are the main indicators of ethnicity?

TESTED

National migration or domestic migration is also important in changing the distribution of population within the UK. For much of the past 50 years, many people have moved from the northern half of Britain to London and the south-east of England. In doing so, they have simply reinforced the UK's uneven distribution pattern (see Figure 5.1).

Over time, the UK population has changed not only its distribution, but also its age structure. This is the product of a number of factors, of which greater **life expectancy** is perhaps the most important.

Life expectancy The average number of years a person born in a particular year might be expected to live

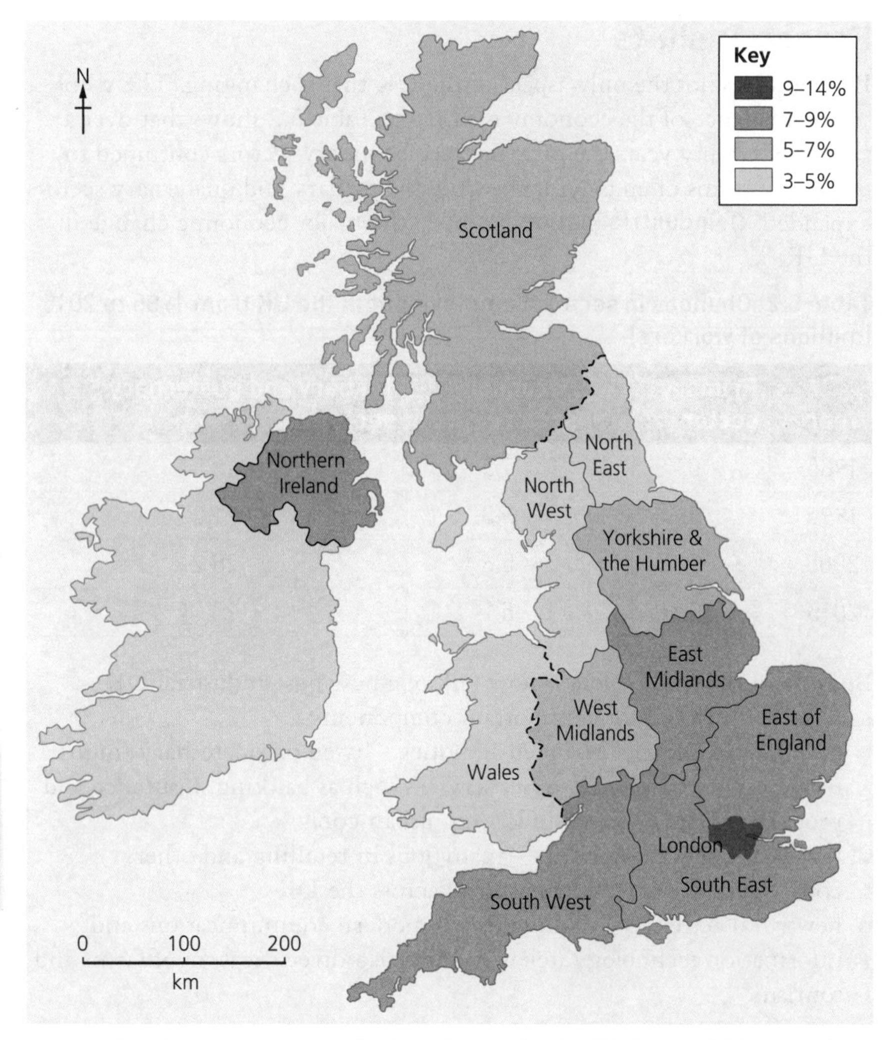

Figure 5.1 Percentage population change in the UK from 2003 to 2013.

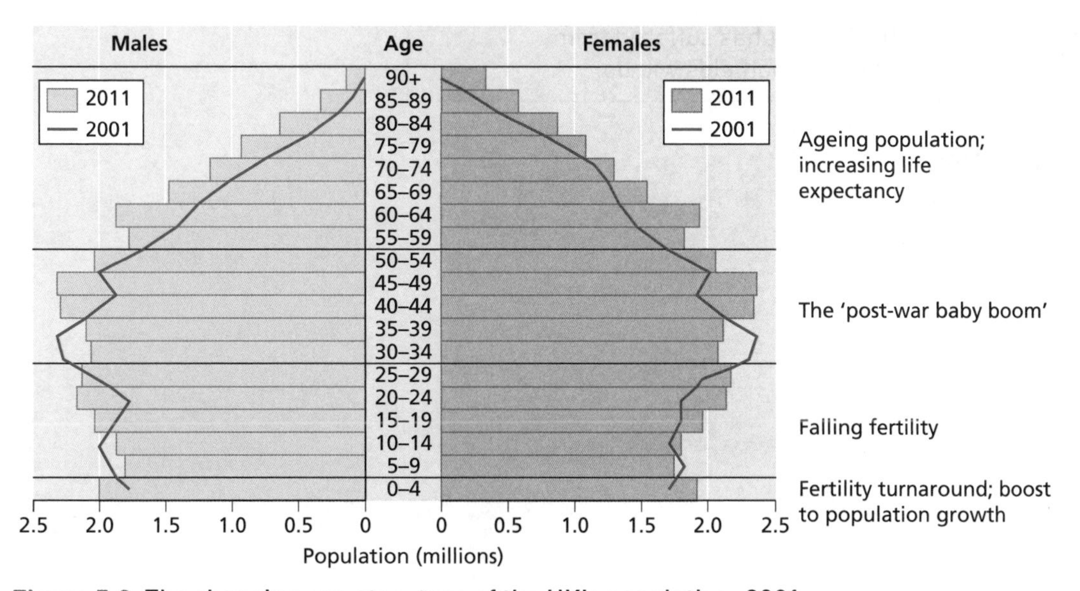

Figure 5.2 The changing age structure of the UK's population, 2001 and 2011.

Now test yourself

TESTED

1. What has persuaded people to migrate southwards in the UK?
2. Why are people today living longer than they were 50 years ago?

Revision activity

Study Figure 5.2 and make brief notes on how the age structure of the UK's population changed between 2001 and 2011.

Economic shifts

Population is not the only aspect of the UK that is changing. The whole sectoral balance of the economy is shifting. Table 5.2 shows that over a period of twenty years the primary and secondary sectors continued to decline in terms of employment, while the tertiary and quaternary sectors expanded. **Deindustrialisation** has been the major economic change in the UK.

Deindustrialisation
The closure of a country's factories due to competition from other countries where raw materials and labour are cheaper and regulations less strict

Table 5.2 Changes in sectoral employment in the UK from 1985 to 2015 (millions of workers)

Year	Primary sector	Secondary sector	Tertiary and quaternary sectors
1985	0.78	7.3	18.3
1995	0.54	6.3	21.0
2005	0.46	5.6	25.2
2015	0.48	5.1	28.1

Because of the shifts, there is now talk of a new post-industrial UK economy which has three important components:

- high-wage, knowledge-based activities – based on quaternary (high-tech) activities and high-order services such as banking, insurance and property; mainly located in leading urban cores
- low-wage, service activities – many jobs in retailing and other commercial services; widely spread across the UK
- new rural activities – making use of modern communications and information technology (teleworking); also diversification of farms and tourism.

Exam tip

Be sure you know of a part of the UK that has suffered from deindustrialisation. Any one of the old coalfields will do.

Now test yourself TESTED

What is meant by teleworking?

Globalisation

Globalisation is a multi-strand process of change. But first and foremost it is about the growth of the **global economy**. Figure 5.3 shows the four main drivers. The promotion of free trade, that is trade without tariffs, is very important. As yet, global free trade does not exist. While being a member of the EU allows UK businesses to trade freely with all of the other member countries, any country outside the EU wishing to trade with the UK or any other member country has to pay tariffs and customs duties.

The unrestricted circulation of capital in the form of **foreign direct investment (FDI)** is also important. As with trade, investment flows are two-way. While UK companies are investing overseas, foreign governments and **transnational corporations (TNCs)** are investing in the UK. The role of the TNCs in the UK has increased with privatisation. This has meant, for example, that services, such as the supply of water, gas and electricity, are no longer run by the government, but by large companies controlled by shareholders.

It is net immigration, the post-industrial economy's trade in services and London's status as a world city that now bind the UK to the global economy.

Globalisation The increased interdependence of countries through economic, social, cultural and political links

Global economy A single economic system resulting from the growing economic links between the countries of the world

Foreign direct investment (FDI) Money from one country invested in another, for example a TNC from the USA building a factory in China

Transnational corporations (TNCs) Companies with operations (factories, offices, warehouses) in more than one country

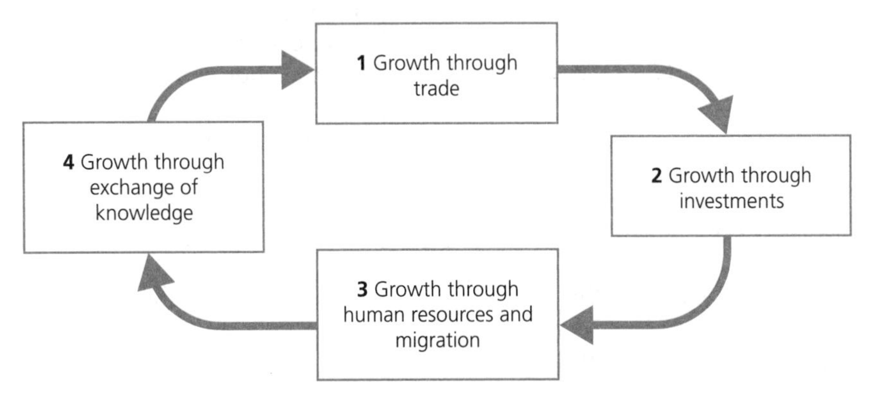

Figure 5.3 The growth of the global economy.

TNCs are powerful players in the global economy. They link national economies in different parts of the world. The top TNCs are involved in three main industries: oil (for example, Shell), electronics (for example, Samsung) and the making of motor vehicles (for example, Honda). Some are specialised in activities such as mining (for example, Rio Tinto) or food and beverages (for example, Nestlé), while others are more broadly based (for example, Mitsubishi).

Now test yourself

TESTED

1. Look at Figure 5.3. What is significant about human resources?
2. Can you name three TNCs, other than the examples given above, that have factories or offices in the UK?

Exam practice

1. Suggest reasons why tourism is thought to be part of globalisation. [4]
2. Assess the reasons for the deindustrialisation of the UK. [8]

ONLINE

How is one major city changing?

The context of the city

REVISED

Site, situation and connectivity

There are three important aspects to a city's location:

- The site: the ground occupied by the city's built-up area. For example, London started life in Roman times as a crossing point of the River Thames. The built-up area has since expanded on both banks of the river (more on the north). It now reaches out to the chalk rim of the syncline known as the London Basin.
- The situation – its location in relation to the locality, region or nation it serves. London's situation is good in terms serving the south-east of England, but it is very much off-centre when it comes to its function as the UK's capital city.
- Connectivity – the most obvious means of connectivity are provided by transport and communications. They not only link the city to those areas it serves, but also provide vital links to the outside world and so facilitate trade, investment, the exchange of information and migration. London is the main focal point in the national rail, motorway and road networks. For connectivity overseas, it possesses the world's second busiest international airport (Heathrow). London Gateway provides the city with all the facilities of large modern port; it is also a global media hotspot.

Now test yourself

What is the difference between situation and connectivity?

TESTED

Exam tip

London is the city chosen to illustrate most of the key ideas and content in this part of the specification. You should have studied one major city in the UK. You should also be prepared to refer to your chosen city in the examination.

City structure

The built-up areas of most towns and cities show the same recurring features (see Figure 5.4):

- a commercial centre (the central business district or CBD)
- industrial estates
- a mix of residential areas – varying in terms of age, dwelling size, house prices, socio-economic status, and so on
- small shopping or service areas
- a fringe of modern 'parks' – retailing, business and science
- green amenity spaces – public parks, sports fields
- schools, colleges and medical services (health centres, hospitals, and so on).

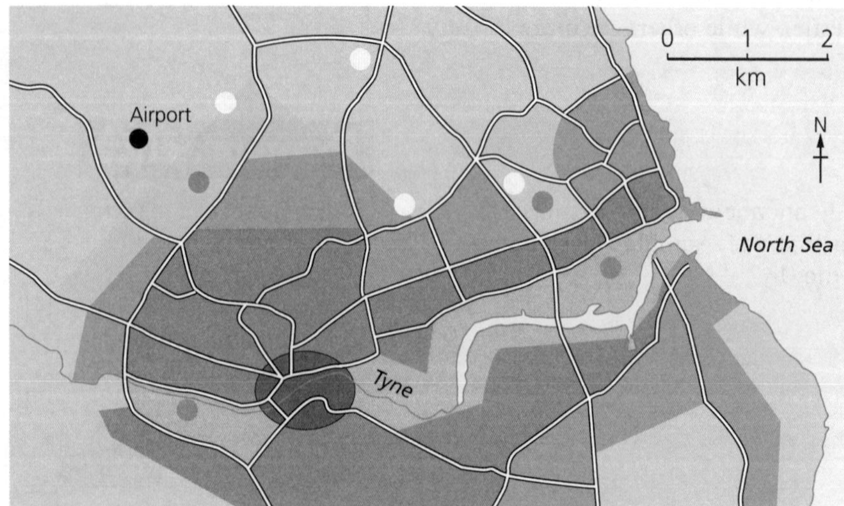

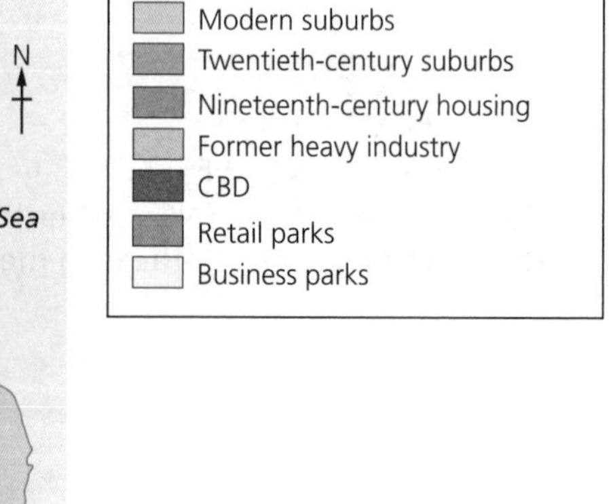

Figure 5.4 The urban pattern of Newcastle upon Tyne.

The main cause of this segregation of land uses is the urban land market. As with the auctioning of any item, a particular piece of land within the built-up area will normally be sold to the highest bidder. The highest bidder will be that activity that can make best use of a site. It is usually retailing that can make the most profitable use of land and property.

Related to this process of bidding is the fact that land values vary within the built-up area. Generally, prices decline outwards from the centre (see Figure 5.5). However, relatively high land values are also found along major roads leading from the centre and around ring roads. Small peaks occur where radial and ring roads cross each other. Businesses will pay extra for sites in these locations because they enjoy good accessibility.

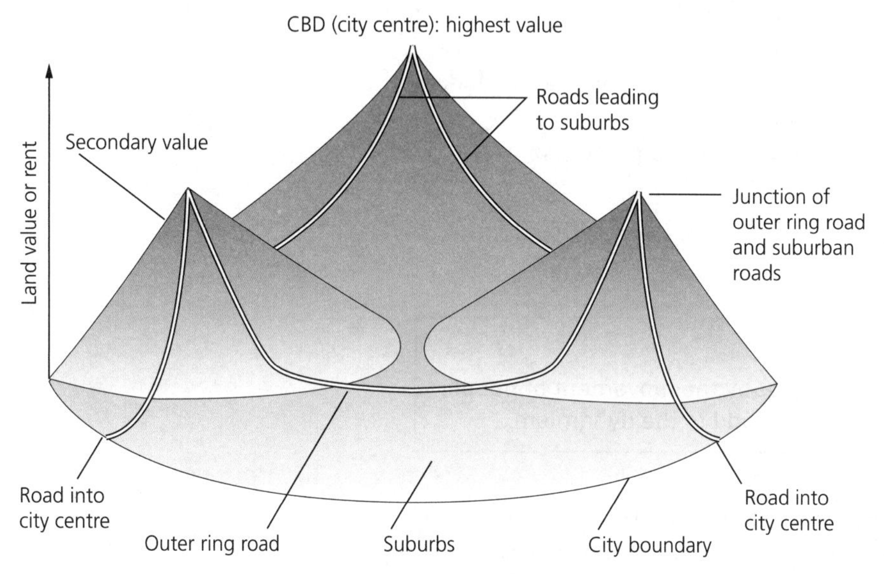

Figure 5.5 Urban land values.

There are other gradients within the city, for example:

- the age of the buildings – this declines from the historic centre
- density – the density of buildings generally declines outwards
- environmental quality – this may improve with declining densities and more modern building.

Because of the importance of distance from the city centre, most cities show a structure of concentric zones (see Figure 5.6).

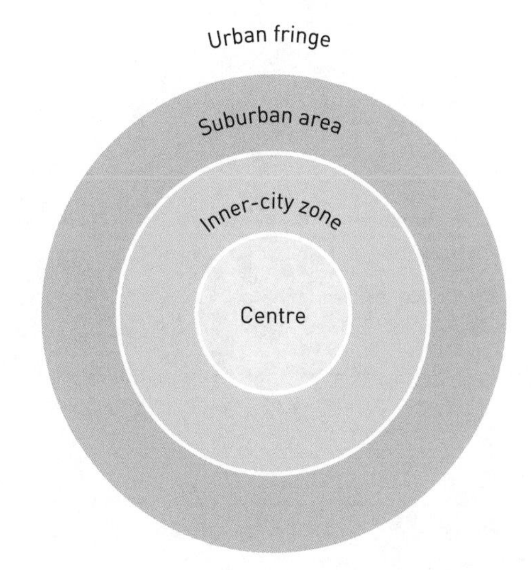

Figure 5.6 Four concentric zones of a UK city.

Now test yourself

1 Look at Figure 5.4. What land uses are found close to the peak land value intersection?
2 What are the attractions of a location on the edge of a city?

TESTED

Exam practice

1 Identify four different types of gradient that exist within a UK city. [4]
2 Study Figure 5.6. State two land uses that are typically found in the urban fringe. [2]

ONLINE

The changing city

REVISED

Cities are dynamic and constantly changing. One of the causes of that dynamism is migration. This is happening at three different scales:

- internationally – people are coming from abroad; at the same time people are emigrating or returning home
- nationally – people are moving in from other parts of the UK; at the same time, people are choosing to move to other cities, towns or rural locations
- internally – people are moving home within the same city.

The causes and impacts of migration

Migration is the result of push and pull factors. In the case of cities, it is the pull factors that are usually stronger. The attractions of cities include:

- more jobs and better employment prospects
- better housing or homes more suited to the changing needs of a growing family
- improved quality of life
- more and better services.

Exam tip

Remember that cities are changed not just by the movement of people. Changes in employment and services also add to the dynamism.

Migration can change many things, from the age structure and ethnic mix of a residential area to pressure on its schools, medical services and housing. One of the most obvious indicators of migration is the concentration of immigrants from ethnic minorities in particular parts of a city. Most recent migrants often seek cheaper rented accommodation in the older and less expensive areas, where large old houses can be subdivided into many single-room flats.

Figure 5.7 shows just how concentrated ethnic minorities are in parts of London. But cheap housing is not the only factor that explains this ethnic segregation. The segregation is also the result of immigrants wishing to live close to family, friends and people from the same ethnic group.

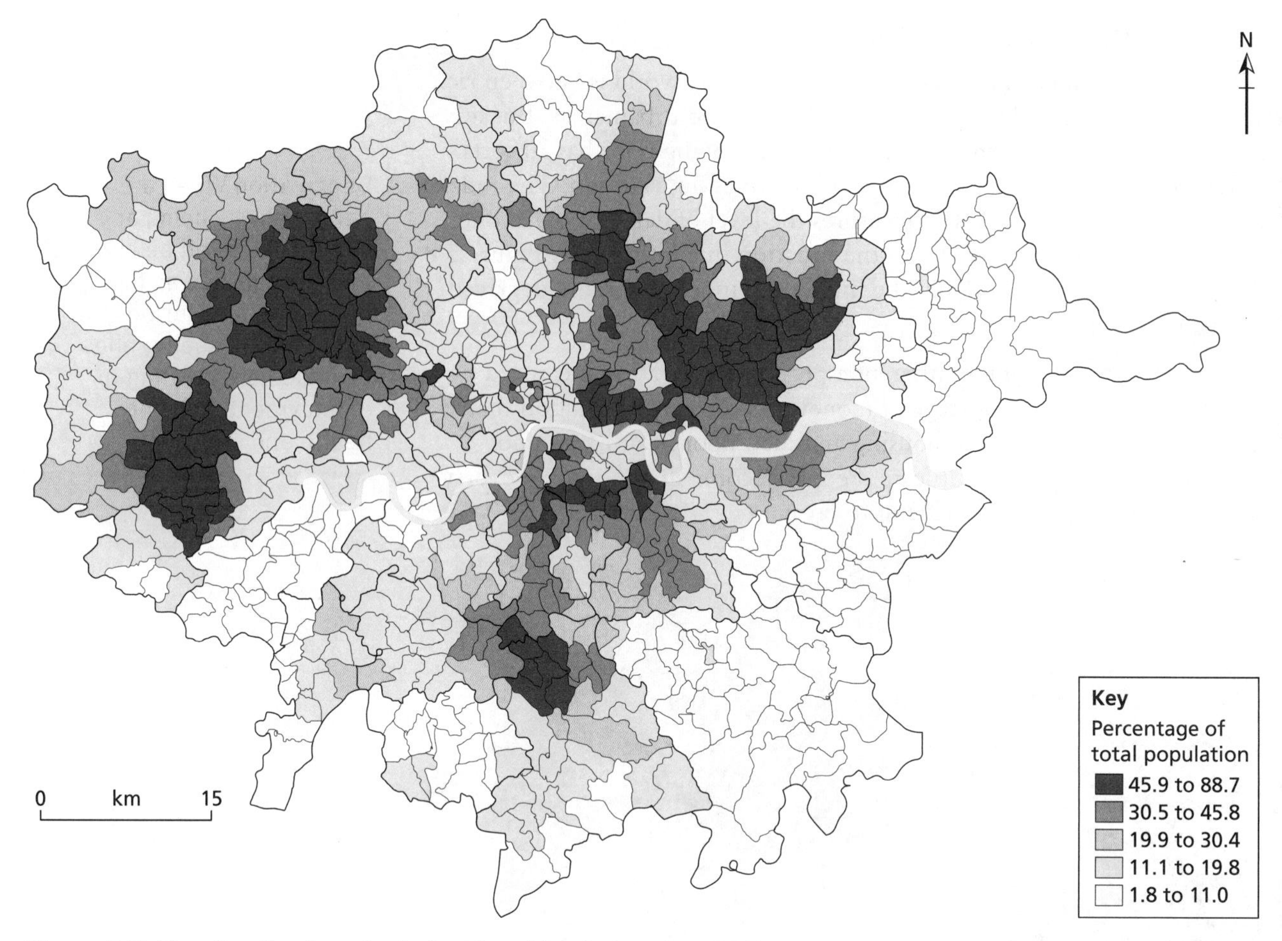

Figure 5.7 The distribution of ethnic minorities in London, 2010.

It is not surprising that different ethnic groups should make their mark on the urban landscapes that they occupy. Table 5.3 shows some of these markers.

Table 5.3 Some ethnic markers in the urban landscape

Marker
Places of worship
Restaurants: ethnic cooking
Grocery shops: ethnic foods
Clothes shops: traditional clothing
Social clubs
Cultural festivals and ceremonies
Cinemas showing ethnic films
Non-English signboards and advertising
Non-English newspapers and magazines

Inequalities within cities

The basic inequality in the UK as elsewhere in the world is between rich and poor. The inequality appears in both rural and urban areas. But it is most marked in cities, especially in core cities. Being poor has all sorts of repercussions which are collectively known as **deprivation**. The index of multiple deprivation (IMD) measures the level of deprivation in one particular area and compares it to the situation in another area. Figure 5.8 shows the situation in London; the higher the index, the more deprived the area.

Deprivation When a person's well-being and quality of life fall below a level regarded as the minimum. Measuring deprivation usually relies on indicators relating to employment, housing, health and education

The basic reason for all the different aspects of deprivation measured by the IMD is income. A shortage of money sets off a chain reaction that includes such things as poor housing and diet, poor health, and poor access to education and healthcare. People become drawn into a poverty trap from which it is very difficult to escape.

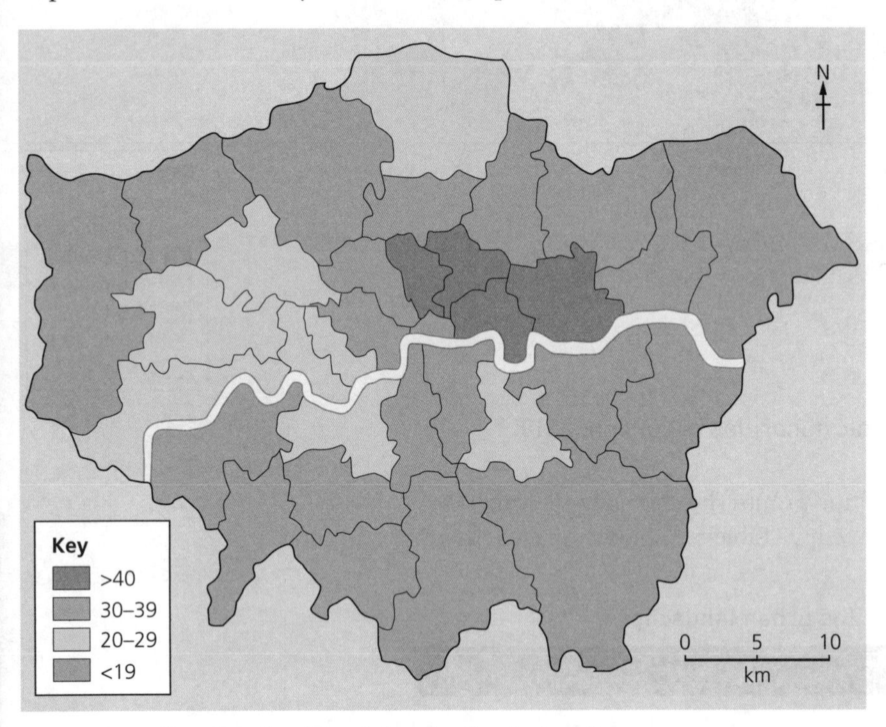

Figure 5.8 **The distribution of deprivation in London.**

Now test yourself

TESTED

Compare Figures 5.7 and 5.8. Is deprivation highest in those parts of London where ethnic minorities are most concentrated?

Exam practice

1 Explain what attracts immigrants to large cities. [4]
2 Give three examples of the inequalities that exist within cities. [3]

ONLINE

City change, challenges and opportunities

REVISED

Change almost inevitably brings challenges and new opportunities.

Decline and decentralisation

London is without doubt a lively and prosperous city. But it has had to face problems in the past, for example a decline of its docks in the 1970s, when cargo began to be carried by new and large container ships. Because of the new ships' sizes they were unable to reach so far up the Thames. A huge area of docks below Tower Bridge soon became derelict. But part of the London Docklands has since been redeveloped into the now famous Canary Wharf – a prosperous mix of banks, offices and fashionable apartments. It is now one of the UK's two financial centres with a global reputation.

Another major change has been the **depopulation** of the older, rundown parts of inner London, such as close to old Port of London. Many families have moved away to the sprawling suburbs. New industrial estates, out-of-town shopping centres, business and science parks on the fringes of London provide proof that it is not only people who have decentralised.

But the **decentralisation** does not end at the fringes of London. For example, many families have moved beyond to **dormitory settlements** in the south-east of England (see Figure 5.9). Some people (and businesses) have or are making the decision to quit London altogether. They have become part of counter-urbanisation.

All this decentralisation has been made possible by improvements in transport.

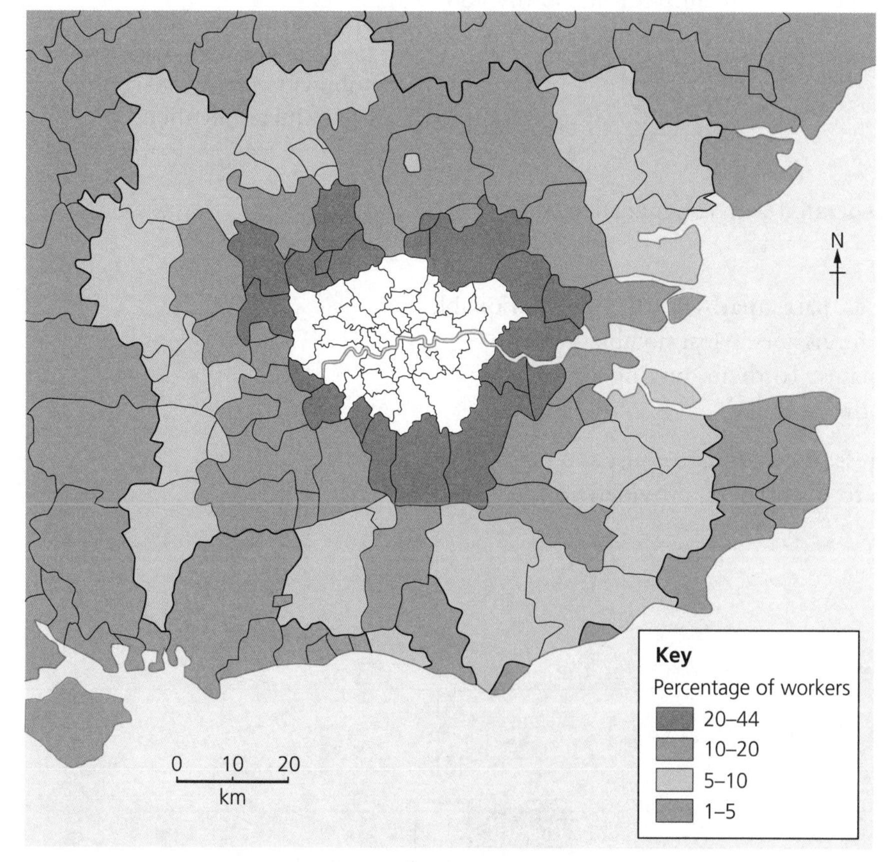

Figure 5.9 Commuting to London.

Depopulation A decline in the number of people living in an area, usually the result of out-migration

Decentralisation The outward movement of people and economic activities from established centres

Dormitory settlement A large residential settlement lying within the commuting area of a city or town functioning as a home base for people who work elsewhere

Now test yourself

What are the advantages to a young family of living in a dormitory settlement rather than in London?

TESTED

But change never stops. Those shopping centres and retail parks of the fringes of London are now being threatened by even newer developments:

- The growth of e-commerce (online shopping) does away with the need for shops.
- The building of two huge shopping centres close to the heart of London, one in Stratford (close to Canary Wharf and the Olympic Park) and the other at Shepherds Bush to the west, mean more jobs once again near the centre.

Regeneration and repopulation

One very noticeable trend in London is the **regeneration** of old residential areas dating from the eighteenth and nineteenth centuries. This is happening because of:

- the cost and stress of commuting long distances into London
- the chance to purchase cheap rundown housing and to modernise it
- the perception that it is 'cool' to be involved in the **gentrification** of an area
- many workers being drawn to London by its global reputation for financial and business services, and as the home of many TNC headquarters. The influx of highly paid workers is creating a huge demand for housing located close to central London.

In some residential areas, gentrification has been preceded by **studentification**.

People involved in this regeneration and repopulation of the inner suburbs benefit in a number of different ways:

- an improved environmental quality
- more leisure time (shorter commutes) to be spent with the family and enjoying the culture and entertainment that London has to offer
- a good return on the money they have spent on their homes.

Regeneration The revival of old and rundown urban areas

Gentrification The movement of high earners into rundown, inner-city residential areas in order to live closer to their place of work. As a consequence, houses and the whole image of an area are improved

Studentification The occupation of areas of cheap housing by students accessible to universities and places of higher education. Such areas are characterised by buy-to-let properties, houses subdivided into one-room flats and small shops

Now test yourself

TESTED

Can you think of any negatives associated with regeneration?

Also helping to boost property prices, particularly in the most fashionable areas, is the investment by wealthy foreigners. Most notable here are Russian tycoons and Arab oil magnates. To them, buying a property in London is a safe haven for some of their money.

So regeneration is reviving an interest in London, not just as a place of work, but as a good place in which to live, that is, provided you have the money!

Exam practice

1 Suggest reasons for building retail and office parks on the fringes of the city. [4]

2 Explain the difference between studentification and gentrification. [4]

ONLINE

Strategies for improving city life

REVISED

Regeneration and rebranding

London has done much in recent years to improve itself. Pride of place must go to east London. Reference has already been made (page 93) to the successful conversion of the London Docklands. The creation of the Queen Elizabeth Olympic Park, where the 2012 Olympic Games were held, has encouraged an impressive upgrading and **rebranding** of nearby residential areas.

Rebranding Changing the image of a place; changing people's perceptions of that place

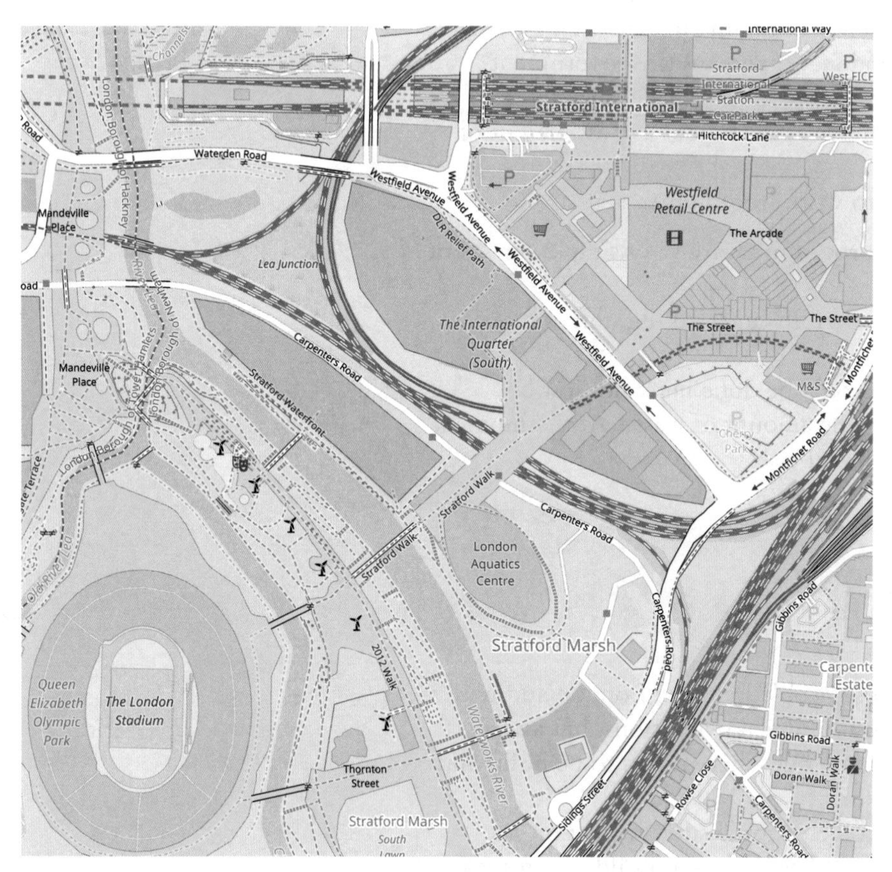

Figure 5.10 The regeneration of East London.

Not so many years ago, this was a huge area of slum housing. Many of the original houses have not been cleared away. Instead, they have been thoroughly gentrified. Table 5.4 reminds us, however, that gentrification does have a downside.

Table 5.4 The positive and negative impacts of gentrification

Positive impacts	Negative impacts
Investment in property improves the slum areas	Restricts the number of low-cost properties for poorer people
Original houses are renovated rather than demolished or left to ruin	Local people might resent new incomers who have no connection to their community
Incoming wealth brings opportunities to local businesses and increases local tax revenues	Existing local services, such as schools and small shops, might not be used by incomers

Now test yourself

Study Figure 5.10. Can you identify the following important elements in the regeneration: the Olympic Stadium, Stratford International Station, Westfield Retail Centre? Why are the last two important?

TESTED

Sustainability and quality of life

Commuting is the activity which threatens the sustainability of London most. Think how much fuel is used every day just to move 3 million people between their homes and places of work in London. Levels of air pollution on some of the streets of London are rising and affecting human health. The quality of life is also being lowered by the stress of travelling on overcrowded trains, buses and roads for several hours each workday.

Solutions to the commuting problem really do not exist. Making greater use of public transport and persuading people to leave their cars at home would be a step in the right direction. But the main issue is employment. There are far too many jobs in London. So perhaps the government should:

- encourage decentralisation and move as many jobs as possible out of London
- encourage those in certain jobs to work from home.

Sustainability can, however, be improved in a variety of simple ways. For example:

- using energy in the home more carefully and making homes more energy efficient
- recycling waste – so much more could be done here, not just by recycling but by simply reducing the amount of waste. At the moment London recycles only a third of its rubbish.

Now test yourself

TESTED

What is meant by sustainability?

Improving the quality of life of those people living in London and within its commuting orbit depends not just on the above actions, but also on:

- providing more affordable housing – but where is this to be built, and where is the money to come from?
- ensuring that green spaces are protected for leisure, recreation and amenity – it would tempting to use these spaces as building sites for affordable housing.

In short, improving the sustainability of London and the quality of life of all its residents is a major challenge with no easy answers.

Exam practice

1 Identify two more impacts to add to Table 5.4: one positive and another negative. [2]

2 Assess possible strategies for making urban living more sustainable. [8]

ONLINE

Urban–rural interdependence

REVISED

Every city has a tributary area, made up largely of rural areas, but also including some towns.

Flows of goods, services and labour

Two-way flows of goods, services and labour represent the interdependence that exists between London and its surrounding rural areas. These rural areas lie within a radius of say 150 km from the capital's centre.

Table 5.5 shows some of the goods and services that are exchanged between London and its accessible rural areas. It is these exchanges that make them interdependent. It is interesting to note that what London has to offer is mainly services, while its tributary rural areas offer more by way of goods.

Table 5.5 The interdependence of London and its rural areas

London to accessible rural areas	Accessible rural areas to London
Range of social services: hospitals, universities	Homes for commuters: mainly in villages
Wide range of employment opportunities	Labour: skilled, unskilled
Cultural venues: museums, art galleries	Fresh farm produce: milk, fruit, vegetables
Entertainment: theatres, cinemas, clubs	Water supply: aquifers in the surrounding chalk
High-order shops	Space for the treatment and disposal of waste
	Open space for recreation and leisure

The exchange has brought economic, social and environmental benefits for both partners. However, when it comes to costs, they fall more heavily on rural areas. They include:

- Economic – local labour is attracted to London jobs, so local employers find it more difficult to fill job vacancies.
- Social – tensions between the incomers (for example, commuters and retirees) and long-term residents of rural settlements. The latter feel that their traditional way of life is being upset. The incomers push up the cost of housing and the locals are priced out.
- Environmental – areas taken over for water supply and wastewater treatment. Disturbance of quiet countryside by increasing numbers of tourists and city people during their leisure time. More traffic and air pollution.

Now test yourself

TESTED

Can you give some examples of the benefits that London and the rural areas around it gain from the interdependence?

Change in a rural area

Case study: Devon

During the last **intercensal period** (2001–2011), the population of this county in south-west England grew by six per cent or 42,000 people. Much of this growth took place in south-east Devon and was the result of new businesses relocating there from more urban parts of the country (see Figure 5.11). Perhaps the best known example was the relocation of the Meteorological Office to near Exeter.

Counter-urbanisation is partly responsible for the continuing increase in population, as well for the new businesses being attracted here by:

- good transport links – air, rail and motorway
- relatively cheap building land
- an attractive part of England in which to live and work
- many workers being easily persuaded to move there
- plenty of local labour here already.

Figure 5.11 shows that much of this counter-urbanisation from other parts of the country ended up in the towns and cities of Devon.

This economic growth and the increase in population are adding to the housing problem. A shortage of affordable housing is the result of:

- a scarcity of building land – much of Devon is protected as a national park and as Areas of Outstanding Natural Beauty (AONBs)
- the price of housing being inflated by the large number of second homes, the many dwellings being let to tourists, and the popularity of Devon as a retirement area.

Traffic congestion and visitor pressure on environmentally sensitive areas are two more problems. These occur because of Devon's accessibility and its fame as a tourist honeypot. The latter is thanks to the county's fine scenery, both inland (Dartmoor National Park) and along its north and south coasts.

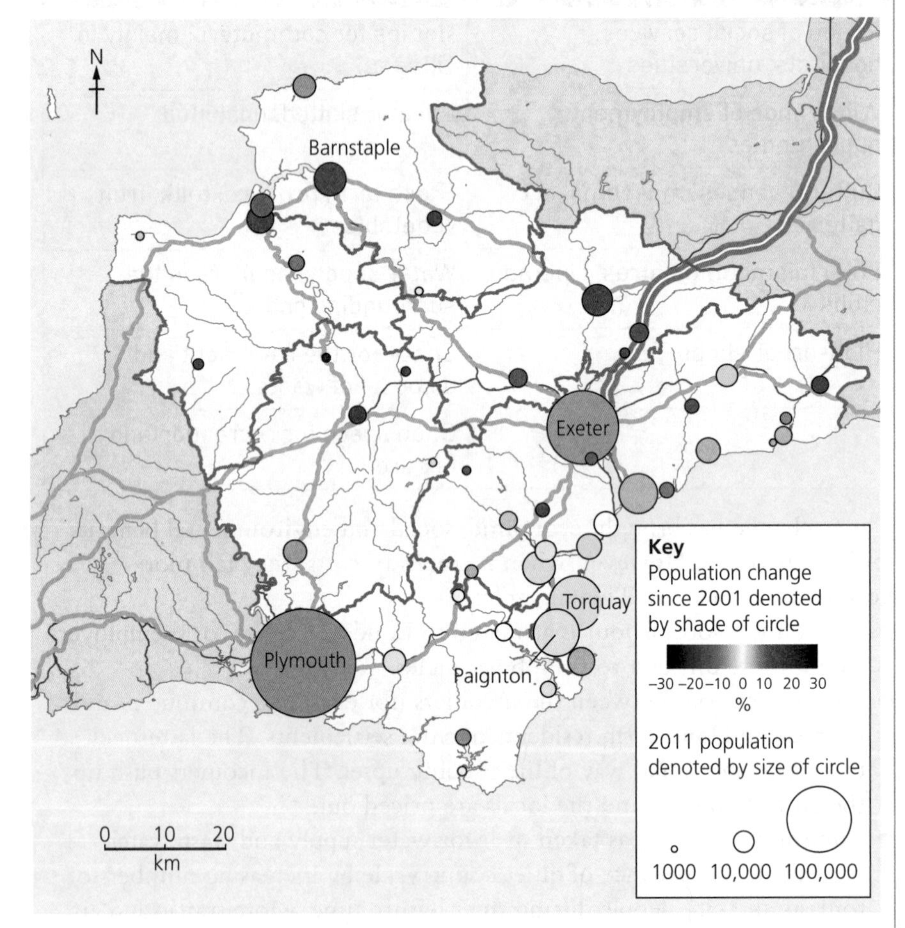

Figure 5.11 Town and city growth in Devon from 2001 to 2011.

Exam tip

The specification asks you to study a named rural area (say a specific county). You should focus on how that area is changing, and why. The case study of Devon is intended to give you some idea of what you should have investigated.

Intercensal period The time between censuses; the census is carried out every ten years in the UK

Now test yourself

TESTED

1 What are the main changes affecting the rural area you have studied?

Exam practice

1 Describe how cities and accessible rural areas are interdependent. [3]
2 Explain the changes that counter-urbanisation brings to rural areas. [4]

ONLINE

Rural change, challenges and opportunities

REVISED

Rural areas, like towns and cities, are constantly changing. In cities, the challenge is how to cope with growth. In rural areas, the opposite is often the case: how to cope with decline is the challenge.

The challenges

- Decline in rural employment – rural jobs today are fewer because farming has become highly mechanised. There are few other jobs in the primary sector.
- Spirals of decline – the lack of job opportunities causes many young adults and their families to move away. They leave behind elderly people and start a spiral of decline of fewer services, increasing isolation and decreasing quality of life (see Figure 5.12).

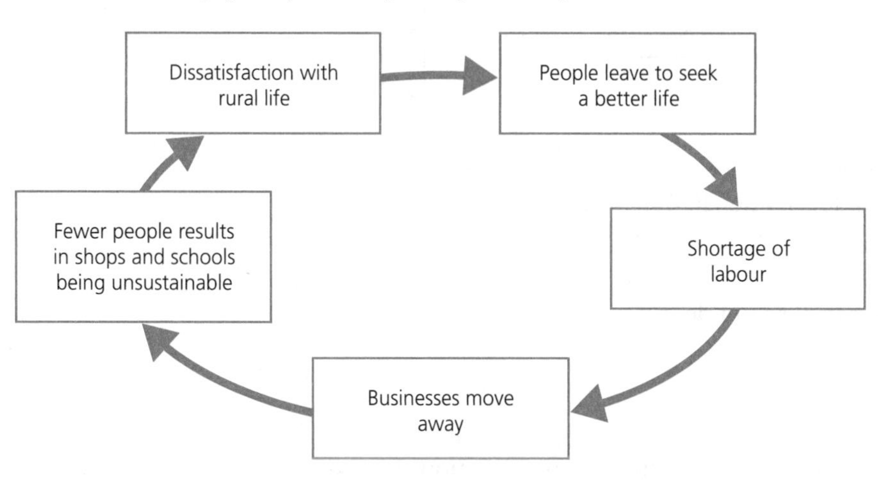

Figure 5.12 The spiral of decline in rural areas.

- Deprivation – is the outcome of the two previous declines. Broadly speaking, the more remote the rural area, the greater the deprivation.

Now test yourself

TESTED

3 What are the main indicators of deprivation?

Now test yourself

2 What other activities are there in the primary sector?

TESTED

Case study: Cornwall

Cornwall is Devon's neighbour, but is more remote. Figure 5.13 shows that much of the county is suffering from high levels of deprivation. Over a third of the villages have no doctor's surgery; many students have to travel over 45 km for sixth-form education.

Much of the deprivation is the result of a large decline in primary employment (farming, fishing, quarrying and mining). Most of the employment is in tourism. But such jobs are poorly paid and are seasonal. This has encouraged out-migration.

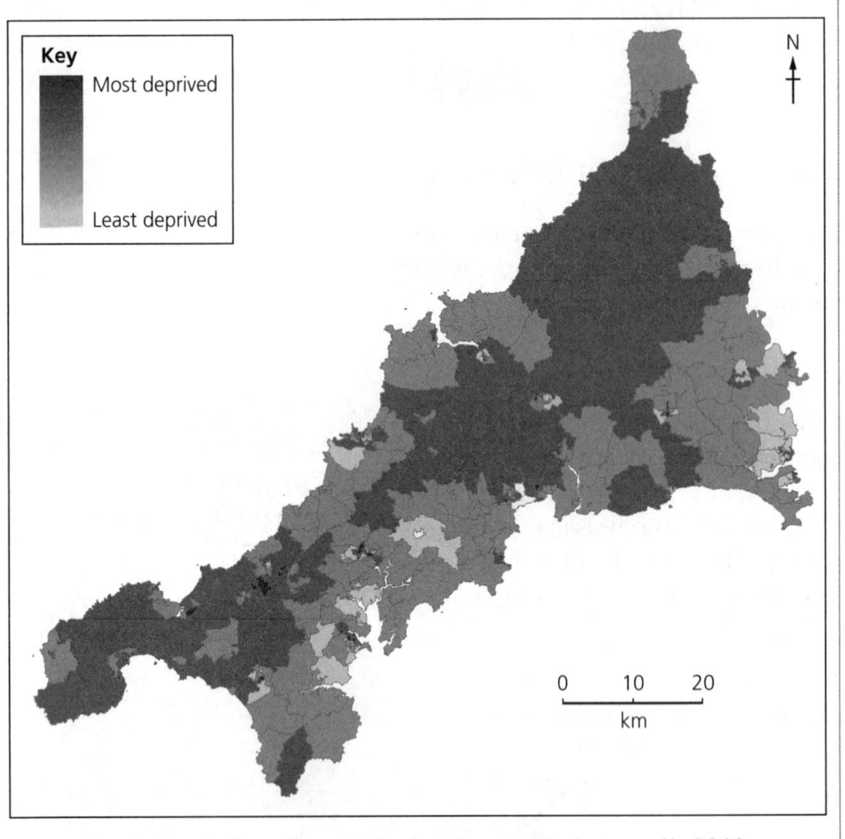

Figure 5.13 The index of multiple deprivation in Cornwall, 2010.

So the challenge in rural areas, like parts of Cornwall, is how to stop the out-migration of people and reduce the level of deprivation.

The opportunities

The answer to the challenge lies mainly in creating jobs and generating more income. This means attracting inward investment and in-migration. Specific opportunities are:

- Farm diversification – many farmers in the UK are finding it difficult to make a living from traditional farming alone. If they want to stay in business, hard-pressed farmers have no choice but to diversify – by doing one of two things:
  - finding other ways of making money out of a farm while continuing to farm, such as becoming an organic farm and setting up their own farm shops, or providing B&B accommodation and farm stays
  - turning the farm into a completely different business, such as a golf course or a pony-trekking centre.

- Recreation, leisure and tourism – these three activities are encouraged in many rural areas by the green open space, attractive scenery and calmer pace of living. Tourism, in particular, creates all sorts of jobs, from work in hotels, guesthouses, cafés and gift shops to staffing tourist attractions. But most of the jobs are part-time and low wage.
- Retirement migration – with many people now expecting to enjoy ten or more years of retirement. As a result, more and more people are moving home once they have retired. More of this retirement needs to be drawn to less accessible rural areas. The arrival of retirees brings more money into the local economy. This, in turn, means more jobs for younger people and an increased demand for services, such as healthcare. Much depends here on making these services accessible to people who perhaps no longer drive a car.
- Teleworking – modern communications, particularly high-speed internet access, mean that an increasing amount of paid work can be done from home. The presence of people of working age earning a good income, plus their children, is good news for the local economy. This also means more support for local services.

While these opportunities promise social and economic benefits, there may be environmental impacts. These might include increased traffic on narrow country roads and building on greenfield sites.

Now test yourself

TESTED

What sort of jobs are suited to teleworking in remoter rural areas?

Exam practice

1 Explain why it is elderly people who suffer most in declining rural areas. [4]

2 Assess the costs and benefits of promoting tourism in rural areas. [8]

ONLINE

Topic 6 Geographical investigations

Revising your fieldwork

For this topic you should have completed two fieldwork investigations:

- one on either coastal environments or rivers
- the other on either urban or rural areas.

Pages 102–5 provide you with some general revision guidance. You should then focus on either coasts (pages 106–7) or rivers (pages 108–9), and urban areas (pages 110–11) or rural areas (pages 112–13) to complete your revision.

The investigation process

REVISED

All geographical fieldwork investigations follow the same process (Figure 6.1). In the exam, you can be tested on your knowledge and understanding of any of the six stages of the investigative process, in two different ways:

- Your own fieldwork investigations, that is, what you did, and what conclusions you came to. This is called a familiar context.
- You can also be asked questions about fieldwork and research methods and data collected by someone else. This is called an unfamiliar context.

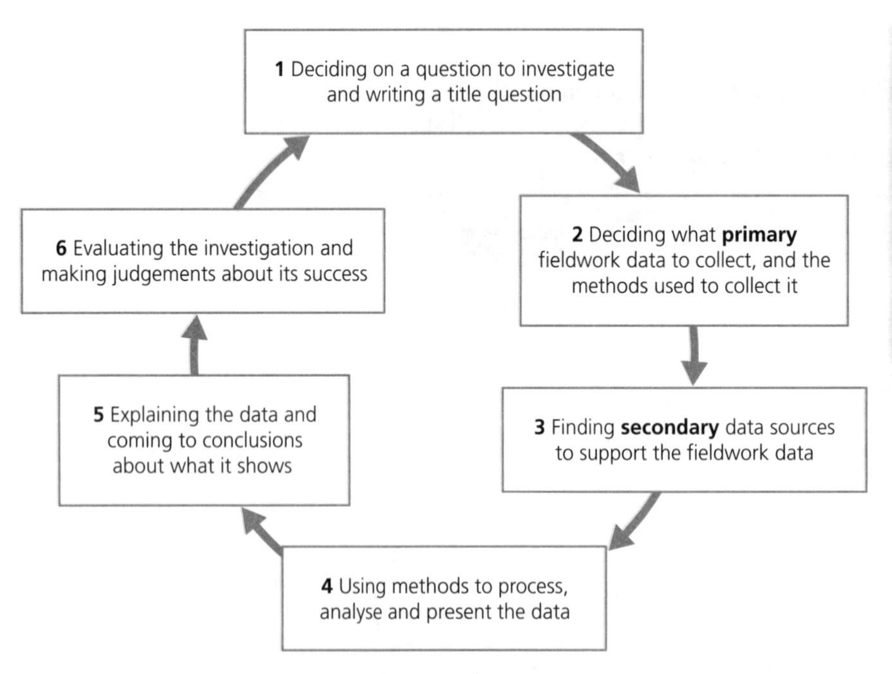

Figure 6.1 The fieldwork investigation process.

Primary data Results you collect yourself in the field: 'first hand' information

Secondary data Any evidence and information you get from another source, such as the internet, maps, articles and books

Revision activity

Look back through your class file and fieldwork notes. Find the enquiry questions or title questions and any key or sub-questions and revise these.

Exam tip

You might have to comment on 'unfamiliar' fieldwork data in the exam, which will often contain errors you need to spot.

Your enquiry questions

An enquiry question is just that, a question. The aim of your fieldwork investigation is to answer it using a combination of **primary data** and **secondary data**. Enquiry questions usually refer to a specific location, such as:

- What have the impacts of new coastal defences at Lyme Bay been on the coastal environment and people?
- How do the characteristics of the River Ouse in York influence flood risk in the city?

In most cases the enquiry question is broken down into two or three smaller key questions. These focus on different parts of the investigation, but by answering all of them you will answer the main enquiry question.

Data collection

REVISED

When designing primary data collection methods you need to be aware of several important considerations:

1. Where should it be collected?
   - The location of the sites chosen to collect fieldwork data is important. Usually more than one site is chosen, to see if one place is different from another. For coast and river data collection five to ten sites are often visited to see if physical processes change along a coastline or down a river.
2. How much should be collected?
   - **Sample size** is important. If you asked only three people a questionnaire survey, or measured only five pebbles in a river or on a beach, your sample will not be representative of the total population of people or pebbles.
3. Will it be accurate?
   - How data collection methods are designed, and which **sampling strategy** is used, has an influence on how likely errors in data collection are and therefore how accurate and reliable your data is.
   - Rivers and coasts fieldwork can involve using specialist equipment (clinometers, callipers, flow meters) and these can be a source of error if not used correctly.

Sample size The number of data points you choose to collect. It is part of data collection design. Sample sizes need to be large enough to be representative of the total population

Sampling strategy Whether random, systematic or stratified data collection was used

Revision activity

Look back at the sites where you collected data. Why were they chosen? Learn their names.

Both primary and secondary data can be of two types (see Table 6.1).

Table 6.1 The two types of primary and secondary date

Quantitative data	Qualitative data
This is number data which directly measures a value, size, length or count	This is data on opinions or views. It includes images, sketches and photographs as well as text
It is often analysed using some maths to work out mean or range	It is very useful, but often harder to analyse than quantitative data

Secondary data

Secondary data may be quantitative, but often it is qualitative. This means there are a number of issues to be aware of:

- Is the secondary data old, or up to date? Often old sources are less relevant, and hard to check for accuracy.
- Is it biased? Many websites, newspapers and other written sources are one-sided and only give one view of an issue.
- Do you know the source? Internet sites, blogs and even online videos can be hard to attribute to a particular person or organisation.

Exam tip

Be as specific as you can when referring to your fieldwork locations. 'Manchester' or 'the Yorkshire coast' is not specific enough!

Revision activity

Make a list of all the secondary sources of data you used. Classify them as quantitative or qualitative. How reliable are they?

Data analysis

REVISED

The quantitative data you have collected, both primary and secondary, should have been used in a process like the one shown in Figure 6.2.

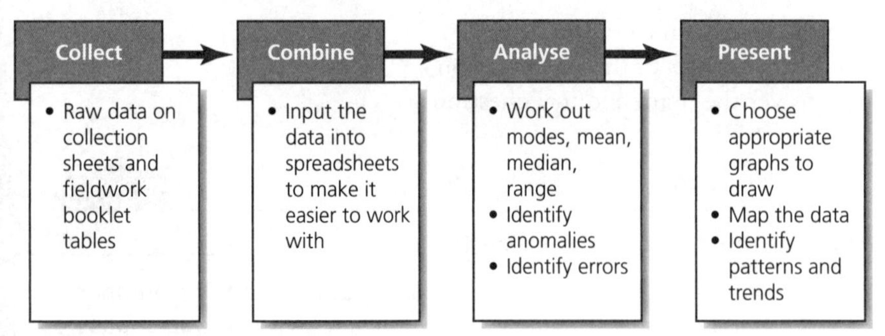

Figure 6.2 Data processing.

In order to be able to understand data and explain it, it needs to be presented in a suitable way. For quantitative data this includes:

- Bar charts, pie charts and bi-polar graphs, which are useful for comparing data sets, sites or locations.
- Tables of data, which can show trends.
- Line graphs, to show how something changes over time.
- Scattergraphs, to show the relationship between two variables, with a best-fit line.
- Maps, to show spatial data and patterns.
- Maps, and other **cartographic** methods, with superimposed graphs, that is combining one type of data with another to show complex patterns.

Qualitative data can be presented by:

- **Annotating** sketches and photographs, to highlight key geographical features.
- Annotating or highlighting text sources, or writing a brief summary of key points.

Cartographic Any presentation technique that uses a paper, or digital, map

Annotating 'Labelling to explain', so using explanatory short sentences rather than single-word labels

GIS

Geographical information systems (GIS) are digital maps that can be altered and changed. In a GIS, such as Google Earth, new layers of information can be added to the digital base map. You can also:

- tag photographs to the digital map
- add pins, to show data collection sites
- add values in the form of graphs, or areas in the form of polygons.

GIS is different from a paper map, because you can change and adapt the GIS repeatedly. GIS can be a powerful tool to both analyse and present your data.

Exam tip

Make sure you learn the names of different types of graphs and what type of data they show.

Revision activity

Look back at the various graphs and maps you completed. Are they accurate and clear, or would another presentation technique have been better?

Conclusions and evaluation

REVISED

The last stages of your investigation should have involved describing, explaining and conclusions (see Figure 6.3).

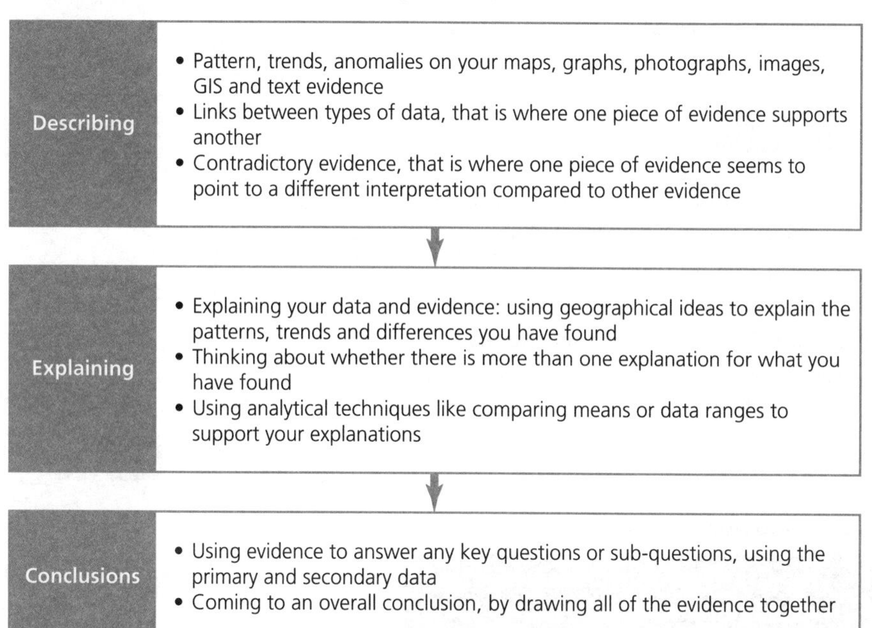

Figure 6.3 Last stages of an investigation.

You should also have evaluated your investigation. This really means standing back, being self-critical and asking how successful the investigation was overall. It includes considering:

- The locations and data collection sites: did you choose the right places, and the right number of places, or would others have been better?
- The design of the fieldwork data collection methods: did you use the right equipment in the right way, or could errors in data collection have occurred?
- Was your sample size large enough: could a small sample size mean that your results might not be very reliable?
- Did you visit at the right time: would your results have been different at a weekend, or when the weather was different?
- The secondary data you used: was it helpful or did it just make the investigating more complex and the conclusions less clear?

You may have found that you are confident about some of your conclusions, but less confident about others. Remember in the exam you can be asked questions about all the stages of the investigative process shown in Figure 6.1 (page 102).

Revision activity

List three reasons why you are not 100 per cent confident about your conclusions, for both of your investigations.

Exam tip

Don't be too self-critical about your fieldwork and make it sound pointless.

Investigating coastal change and conflict

The focus of this investigation is the impact of coastal management on coastal processes and communities.

Enquiry questions and location

REVISED

- This enquiry is about investigating how coastal management methods like groynes and sea walls, or strategies such as 'do nothing' have affected the physical processes and people at the coast.
- It could be investigated in places with rapid coastal erosion where there are conflicting views about how a coast should be managed, such as Holderness, north Norfolk or Christchurch Bay.

Fieldwork methods

REVISED

Table 6.2 Coastal change and conflict fieldwork methods

Quantitative fieldwork on beach morphology and sediment characteristics	Qualitative fieldwork on coastal management measures and their success
Beach profiles are measured to determine the slope of the beach and any features like berms (Figure 6.4) A sample of 20–30 pebbles is measured using a ruler (*a*-, *b*- and *c*-axes) and shape measured using a roundness scale Similar measurements can be done along a beach, to see if groynes have altered longshore drift	Photographs and field sketches of coastal defences and the impact of strategies like 'do nothing' Written descriptions of sites, changes and impacts like damaged property or inaccessible areas

Beach profiles The changing shape, slope angle, sediment size and roundness across the width of a beach

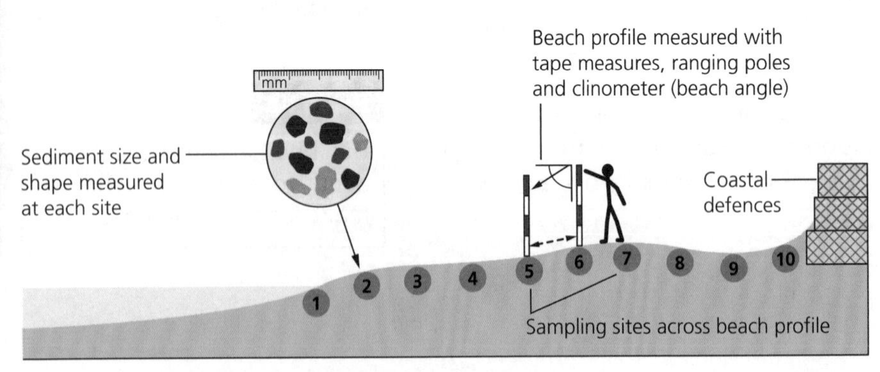

Figure 6.4 Measuring a beach profile.

Secondary data

REVISED

You should have used a geology map. These are sourced from the British Geological Survey (BGS) and show:

- Solid geology, that is sedimentary, igneous and metamorphic rocks at the coast – some of which are resistant to erosion but others less so. It also shows faults and other geological weaknesses.
- Drift geology, that is sediments such as boulder clay and sand, which are easily eroded by the sea.

It is a reliable source, but you may need to visit a site to confirm the geology shown on the map or spot any local differences.

Another useful source of secondary data is the local **shoreline management plan (SMP)**. It includes areas of 'hold the line' and 'do nothing' as well as types of coastal defence, and an explanation of why decisions about protection have been made.

Shoreline management plan (SMP) A planning document that sets out how a coast is going to be protected – or not

Exam practice

1 Explain one way a secondary data source you used supported your investigation. [2]
2 Explain one possible source of error with the method you used to measure a beach profile. [2]
3 Explain how the fieldwork data collection you completed supported your conclusions. [4]
4 Evaluate the methods you chose to analyse and present the primary fieldwork data that you collected. [8]

ONLINE

Revision activity

Make revision notes on your own fieldwork methods, including any problems you encountered and sources of errors.

Exam tip

Try and use the right equipment names: a 'clinometer' not 'the thingy that measures angles'.

Exam tip: familiar and unfamiliar questions

In Paper 2 there will be questions to answer on fieldwork. These questions will be of two types, familiar and unfamiliar.

Familiar questions:

- These test your understanding of the fieldwork you completed.
- They will include a phrase like '*You have carried out a fieldwork investigation in a coastal environment.*'
- The use of 'you' lets you know that these questions are about your own fieldwork.

Unfamiliar questions:

- These test your wider understanding of fieldwork.
- You will be asked to look at some fieldwork data or information, such as a figure.
- These questions will use a phrase like '*A group of students collected information about quality of life.*'
- This lets you know the question is about fieldwork by someone else.
- The questions will ask you to assess their fieldwork, that is, consider their methods, presentation or conclusions in terms of their success.

Investigating river processes and pressures

The focus of this investigation is how and why drainage basin characteristics influence flood risk for people and property.

Enquiry questions and location

REVISED

- This investigation might look at land use (urban, farmland, woodland) in a drainage basin, as well as factors such as geology (permeable, impermeable), valley gradient and river management such as existing flood defences. All can affect flood risk.
- A small part of a drainage basin is often studied, perhaps including residential areas at risk from flooding and areas that have recently been flooded, for example, the River Severn, the Ouse in York or Valency in Boscastle.

Fieldwork methods

REVISED

Table 6.3 River processes and pressures fieldwork methods

Quantitative fieldwork on changes in river channel characteristics	Qualitative fieldwork on factors that influence flood risk
At each site, the river channel profile is measured in terms of width and depth (Figure 6.5) A flow meter can be used to measure water velocity A sample of 20–30 pebbles is measured using a ruler (*a*-, *b*- and *c*-axes) and shape measured using a roundness scale River gradient can be measured at each site using a clinometer	Photographs and field sketches of river features and flood defences, and areas at risk from flooding Written descriptions of sites, changes and impacts like damaged property or inaccessible areas

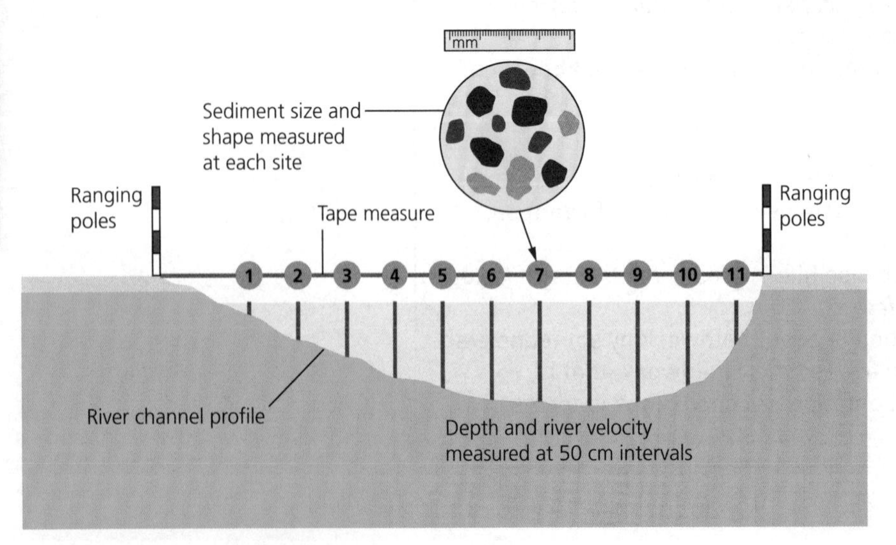

Figure 6.5 Measuring channel characteristics.

Secondary data

REVISED

The Environment Agency produces flood risk maps which you should have used:

- These identify areas of high, medium and low flood risk.
- They are used by people and businesses to assess if their property is at risk.
- They also help decide if flood defences should be built or improved.

Flood risk maps are not perfect. They are likely to have to change as global warming increases risk or land use changes, such as building more urban areas.

Another useful source is historic river discharge data. This shows how rivers reacted to heavy rainfall events in the past, and is usually in the form of a **hydrograph**.

Hydrograph A graph that shows how the volume of water in a river responds to rainfall, and identifies periods of flooding

Exam practice

1 Explain one way a secondary data source you used supported your investigation. [2]
2 Explain one possible source of error with the method you used to measure river channel depth. [2]
3 Explain how the fieldwork data collection you completed supported your conclusions. [4]
4 Evaluate the methods you chose to analyse and present the primary fieldwork data you collected. [8]

ONLINE

Revision activity

Make revision notes on your own fieldwork methods, including any problems you encountered and sources of errors.

Exam tip

Try and use the right equipment names: a 'clinometer' not 'the thingy that measures angles'.

Exam tip: fieldwork 'evaluate' questions

One of the 8-mark fieldwork questions on Paper 2 will be a 'familiar' question about your own fieldwork:

- this will use the command word 'evaluate'
- evaluate means measuring or judging the value or success of something
- you should consider the strengths and weaknesses of your own fieldwork
- you will need to include a conclusion, backed up by evidence
- this question could focus on the planning stage of your fieldwork, the data collection methods, the presentation methods you used, the way you analysed your data or the conclusions you came to.

Investigating dynamic urban areas

The focus of this investigation is how and why quality of life varies within urban areas.

Enquiry questions and location

REVISED

- This question is best answered by comparing two or more areas. These could be close to each other in a city, or a comparison between areas in different towns or cities.
- The areas chosen need to be small, such as urban wards or **census** LSOAs (Lower Layer Super Output Areas) in order to make a detailed comparison.
- The focus is on quality of life, which includes environmental quality (green space, pollution, crime), access to services (shops, entertainment), housing quality (size, age, repair) and employment.

Fieldwork methods

REVISED

Census A government survey carried out every ten years. All households complete a long questionnaire about their lives

RICEPOTS A way of classifying land use when completing a land-use map: Residential, Industrial, Commercial, Entertainment, Public, Open space, Transport, Services

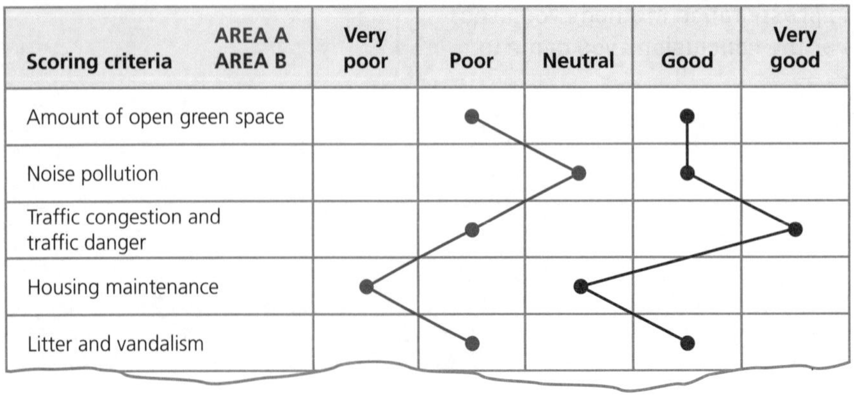

Table 6.4 Dynamic urban areas fieldwork methods

Quantitative fieldwork on environmental quality	Qualitative fieldwork on the views and perceptions of quality of life
Environmental quality surveys, using a bi-polar scoring system (Figure 6.6) to make judgements about environmental quality Questionnaire surveys to local people and/or visitors using closed questions and Likert scales for ease of analysis Land-use maps using the **RICEPOTS** system and a base map, to record what the areas are used for	Interviews with local residents and with local managers (the council, estate agents, community group leaders) are a very good way to get detailed insights into places Photographs can be used to produce comparative montages of the areas

Scoring criteria AREA A AREA B	Very poor	Poor	Neutral	Good	Very good
Amount of open green space					
Noise pollution					
Traffic congestion and traffic danger					
Housing maintenance					
Litter and vandalism					

Figure 6.6 Part of a bi-polar environmental quality survey.

Secondary data

REVISED

You should have used the UK national census data from the Office for National Statistics. This is a useful source of information on:

- economic activity: types of job, employment and unemployment
- households: size of households, housing type, education levels
- population: age, ethnicity and gender.

However:

- The census only happens once every ten years, so the data may be out of date.
- The census provides average data for small areas, but not for individual streets or estates.
- Some useful data, like income levels, is not included in the census data.

Another useful source of secondary data is the index of multiple **deprivation** (IMD). It is very useful for comparing quality of life between small areas.

Deprivation Lacking something, like a decent income or safe, warm housing, that the rest of society considers essential

Exam practice

1 Explain one problem with a secondary data source you used to support your investigation. [2]

2 Explain one fieldwork method you used to determine quality of life. [2]

3 Explain why you chose your fieldwork data collection sites. [4]

4 Evaluate the success of the primary data collection methods used in your investigation. [8]

ONLINE

Revision activity

Make a list of the secondary data sources you used in both investigations, including details of where they were from, how old they were and their reliability.

Exam tip

Make sure you state what specific information you extracted from the census or other secondary sources.

Exam tip: familiar and unfamiliar questions

In Paper 2 there will be questions to answer on fieldwork. These questions will be of two types, familiar and unfamiliar.

Familiar questions:

- These test your understanding of the fieldwork you completed.
- They will include a phrase like '*You have carried out a fieldwork investigation in a coastal environment.*'
- The use of 'you' lets you know that these questions are about your own fieldwork.

Unfamiliar questions:

- These test your wider understanding of fieldwork.
- You will be asked to look at some fieldwork data or information, such as a figure.
- These questions will use a phrase like '*A group of students collected information about quality of life.*'
- This lets you know the question is about fieldwork by someone else.
- The questions will ask you to assess their fieldwork, that is, consider their methods, presentation or conclusions in terms of their success.

Investigating changing rural areas

The focus of this investigation is how and why deprivation varies within rural areas in the UK.

Enquiry questions and location

REVISED

- This question is best answered by comparing two or more areas. These could be two places in the same rural area, perhaps one that is growing versus another that is declining.
- The areas chosen need to be small, such as villages (honeypot versus a commuter village) in order to make a detailed comparison.
- The focus is on quality of life, which includes environmental quality (green space, pollution, crime), access to services (shops, entertainment), housing quality (size, age, repair) and employment.

Fieldwork methods

REVISED

Table 6.5 Changing rural areas fieldwork methods

Quantitative fieldwork on environmental quality	Qualitative fieldwork on the views and perceptions on quality of rural life
Environmental quality surveys, using a bi-polar scoring system to make judgements about environmental quality Questionnaire surveys to local people and/or visitors using closed questions and Likert scales (Figure 6.7) for ease of analysis Service counts, to determine the range of services available in contrasting villages	Interviews with local residents and with local managers (farmers, estate agents, community group leaders, local businesses) are a very good way to get detailed insights into places Photographs can be used to produce comparative montages of the areas

Q3 Has the closure of rural services in the village affected you?

Yes ☐ No ☐ Don't know ☐

Q4 This village has become a better place to live over the last 10 years:

Strongly agree ☐	Agree ☐	Neutral ☐	Disagree ☐	Strongly disagree ☐

Q5 Tick the **two** services you would most like to have available in the village:

Pub ☐	Mobile library ☐	Youth club ☐
Café ☐	Sunday bus services ☐	Doctor ☐

Figure 6.7 **Part of a questionnaire using closed questions and a Likert scale (Q4).**

Secondary data

REVISED

You should have used the UK national census data from the Office for National Statistics. This is a useful source of information on:

- economic activity: types of job, employment and unemployment
- households: size of households, housing type, education levels
- population: age, ethnicity and gender.

However:

- The census only happens once every ten years, so the data may be out of date.
- The census provides average data for small areas, but not for individual streets or estates.
- Some useful data, like income levels, is not included in the census data.

Another useful source of secondary data is the index of multiple deprivation (IMD). It is very useful for comparing quality of life between small areas.

Revision activity

Make a list of the secondary data sources you used in both investigations, including details of where they were from, how old they were and their reliability.

Exam practice

1 Explain one problem with a secondary data source you used to support your investigation. [2]
2 Explain one fieldwork method you used to determine quality of life. [2]
3 Explain why you chose your fieldwork data collection sites. [4]
4 Evaluate the success of the primary data collection methods used in your investigation. [8]

ONLINE

Exam tip

Make sure you state what specific information you extracted from the census or other secondary sources.

Exam tip: fieldwork 'evaluate' questions

One of the 8-mark fieldwork questions on Paper 2 will be a 'familiar' question about your own fieldwork:

- this will use the command word 'evaluate'
- evaluate means measuring or judging the value or success of something
- you should consider the strengths and weaknesses of your own fieldwork
- you will need to include a conclusion, backed up by evidence
- this question could focus on the planning stage of your fieldwork, the data collection methods, the presentation methods you used, the way you analysed your data or the conclusions you came to.

Topic 7 People and the biosphere

Why is the biosphere so important and how do humans use its resources?

Biomes and their distribution

REVISED

The Earth's surface is covered by a thin layer of living material called the biosphere, made up of **ecosystems**. On a global scale these ecosystems are referred to as **biomes**. Figure 7.1 shows the global distribution of biomes.

The pattern of biomes is strongly influenced by the Earth's climate:

- Tropical forests grow in equatorial regions, where temperatures are over 28 °C all year round and there is almost continual precipitation from convection where the Hadley cells meet.
- Temperate forests, like the UK's, are found 40–55° north, in an area of high rainfall where the Polar and Ferrel cells meet, but with very seasonal temperatures.
- The northern taiga (boreal or coniferous) forests are found much further north, and are adapted to cope with extreme winter cold and lack of available water.

Ecosystem A community of plants and animals interacting with the non-living environment (rocks, the atmosphere, water and soil)

Biome A global scale ecosystem, such as tropical forest, which has similar characteristics but is found in a number of different locations

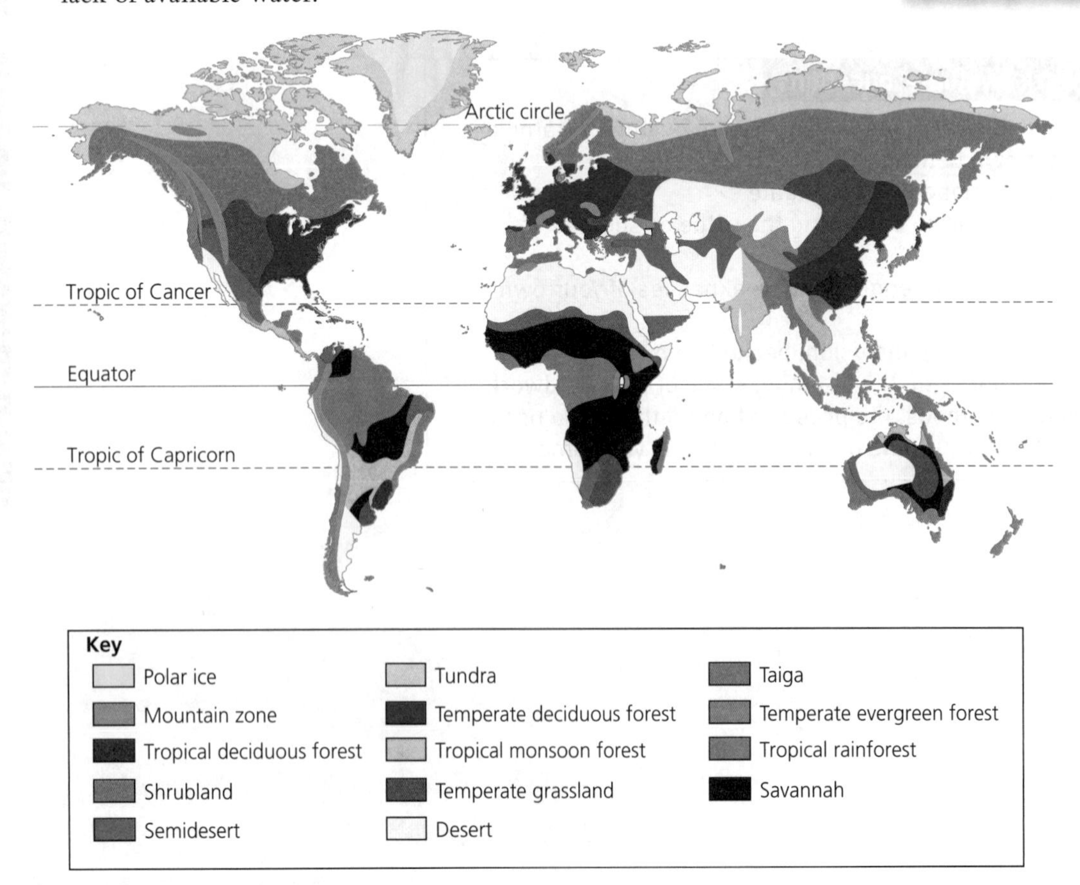

Key

Polar ice	Tundra	Taiga
Mountain zone	Temperate deciduous forest	Temperate evergreen forest
Tropical deciduous forest	Tropical monsoon forest	Tropical rainforest
Shrubland	Temperate grassland	Savannah
Semidesert	Desert	

Figure 7.1 Global distribution of biomes.

- Deserts encircle the Earth 25–35° north and south, at the descending (high-pressure) arm of the Hadley cell where there is very low rainfall (<400 mm/year) and extremely high temperatures and long sunshine hours.
- Grasslands, such the tropical savannahs and temperate prairies and steppes, are found in places with highly seasonal rainfall, which in total is too low for tree and forest growth.
- The tundra is a cold, treeless grassland biome found in the Arctic where both temperatures and rainfall are too low for tree growth and the growing season is only three to five months long.

Local factors

The global climate system and atmospheric circulation you revised in Topic 1 is the main factor that controls which biomes are found where. This is because:

- temperature falls the further north or south from the equator
- **seasonality** increases with distance from the equator
- precipitation is related to low-pressure (high) versus high-pressure areas (low).

However, there are a number of local factors that can alter the distribution of biomes. This means:

- some biomes can be found in unexpected locations
- biomes contain different mixes of plant and animal species.

One important factor is continentality. The further from the ocean a place is, generally the less precipitation it receives. This means the interior areas of Asia and North America are dry compared to their coasts. Figure 7.2 shows the influence of local factors.

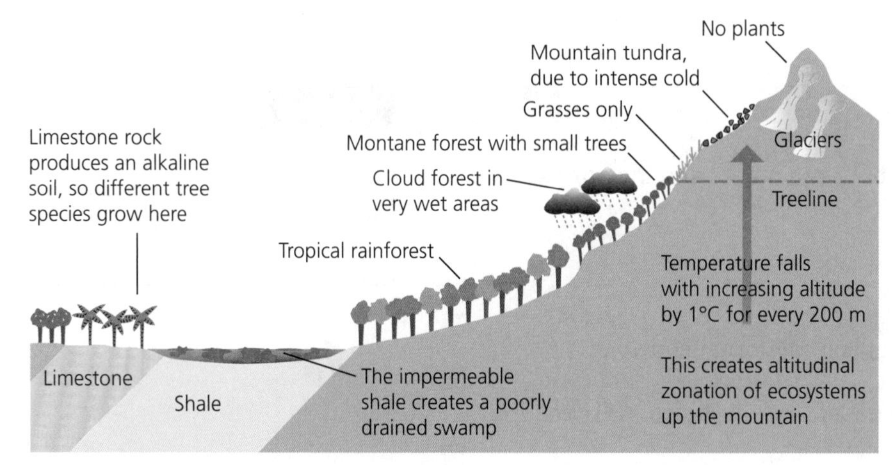

Figure 7.2 Local factors and biomes.

Now test yourself

TESTED

How much does temperature fall for every 200 m increase in altitude?

Exam tip

Make sure you can roughly draw the distribution of tropical forests and taiga on a blank world map.

Revision activity

Look back at Topic 1 and review the global atmospheric circulation. Try to link the pattern of atmospheric cells to the pattern of biomes.

Seasonality The variability of precipitation and temperature over the course of a year

Revision activity

Draw a spider diagram summarising all the local factors that can affect biome distribution.

Interactions

All ecosystems and biomes are complex systems of interactions between their **abiotic** and **biotic** components. All components are linked by processes, all of which need to operate normally to keep the ecosystem healthy (see Figure 7.3). The most important processes are:

- photosynthesis by plants, that fix energy from the Sun into carbohydrate (new plant growth)
- food webs, that move energy through the ecosystem from plants to herbivorous and carnivorous animals
- nutrient cycling: plants take up nutrients from the soil via their roots, which then move along food webs, and are returned to the soil by decay
- the exchange of gases, oxygen and carbon dioxide between plants and animals, and the atmosphere.

Abiotic The non-living parts of an ecosystem: rock, water, the atmosphere, soil

Biotic Any living part of an ecosystem, such as plants (flora), animals (fauna), bacteria and fungi

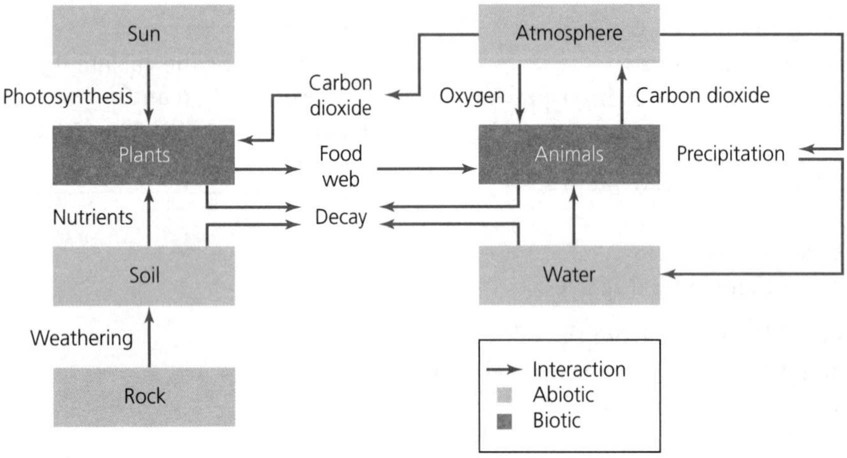

Figure 7.3 A simplified ecosystem and processes.

Now test yourself

TESTED

Name the two gases exchanged between plants and the atmosphere.

Exam tip

Try and use the correct terminology when explaining ecosystems and their processes.

Exam practice

1 Describe the distribution of tropical rainforest shown in Figure 7.1 on page 114. [2]
2 Explain the distribution of the desert biome shown in Figure 7.1 on page 114. [4]
3 Explain how local factors can alter biome distribution. [4]
4 Explain two interactions between the biotic and abiotic parts of ecosystems. [4]

ONLINE

A life-support system

REVISED

The biosphere and its biomes provide people with a wide range of resources which are called **goods**. The small number of remaining **indigenous people** who live within tropical forests and on tropical savannahs have the most direct relationship with ecosystems and use forests and grasslands to obtain many resources:

- Foods, which can be hunted and gathered, including animals (called 'bushmeat') and fish; the roots of many plants are edible and a good source of carbohydrate, and tropical fruits are a source of vitamins and sugars.
- Traditional medicines obtained from plants, tree-bark and roots.
- Building materials such as trees for timber and vines to make rope and lashings.
- Fuelwood from trees, either burnt directly or used to make charcoal.

In many places, local people who have turned to farming as a way of life still use forests for fuelwood. In Africa, there is a huge trade in bushmeat. Animals such as monkeys, lemurs and antelope are hunted, often smoked to preserve the meat, and then traded long distances to large cities.

This indigenous and local resource use has a limited impact on ecosystems, although over-hunting can threaten species with extinction.

Commercial use of forest resources is a much greater threat (see Table 7.1).

Table 7.1 Commercial use of forest resources

Energy resources	Water resources	Mineral resources
Ecosystems often have fossil fuels beneath them Commercial oil and gas drilling leads to forest destruction There is also a risk of oil spills and even fires breaking out Forests are increasingly cut down to grow biofuel crops such as palm oil	In tropical areas, large rivers and high rainfall make some places ideal for hydroelectric power This involves constructing large dams, and flooding vast forest areas to create a water reservoir Indigenous people are often displaced, as well as forests being destroyed	Mining for minerals such as gold, coltan (used in your mobile phone) or iron ore is widespread in many forests This leads to deforestation Worse, waste from mining often pollutes rivers and lakes and the pollution can be carried long distances

Both tropical forests and the northern coniferous taiga forests are threatened by commercial resource exploitation.

Goods Physical resources which humans extract from ecosystems to use, like foods and timber for building and fuel

Indigenous people The original inhabitants of an area; some still live in traditional tribes, but the number is declining

Commercial For profit, and often at a large scale

Revision activity

Make a quick list of activities you have done today that have used energy, water and mineral resources.

Biosphere regulation

As well as goods, the biosphere provides a range of important **services** which help to keep the planet as a whole healthy. These functions are often called regulating services. They include links between the biosphere and other physical systems (see Table 7.2).

Table 7.2 Regulating services

Composition of the atmosphere	Soil health	Hydrological cycle regulation
Plants **sequester** carbon dioxide from the atmosphere as they grow, but release it back when they die and decay Forests in particular are important carbon stores Destroying forests releases carbon dioxide and prevents it being sequestered In addition, plants put oxygen back into the atmosphere	Healthy soils are ones which contain **nutrients** Plants remove nutrients from the soil as they grow, which are then passed through the ecosystem food web Dead plant and animal matter, called litter, then decays and returns nutrients to the soil Breaking this cycle, by deforestation for instance, leads to declining soil health	Plants, especially forests, intercept falling precipitation, slowing it down Plants also absorb water through their roots, and prevent surface runoff by slowing down flowing water This prevents flooding and regulates river flow Plants return water to the atmosphere by transpiration This contributes to the formation of clouds and rain

Services Physical processes within biomes that help to maintain bigger physical systems like the hydrological (water) cycle and atmospheric system

Sequester To take in and store

Nutrients Chemical elements and compounds, such as phosphate and potassium, that plants need to grow

Affluence Increasing wealth

Industrialisation The transition from a farming economy to one where people work in secondary (factory) and tertiary (office) jobs

Resource demand

The 7.4 billion world population in 2017 already has a huge demand for land, food, water and energy resources. Projections of future population suggest there will be 9.5 billion in 2050, and 10.8 billion by 2100.

Population growth is not expected to be even across different global regions, as Figure 7.4 shows. It is not just the total number of people that will lead to increased resource demand:

- Since 2007, 50 per cent of people have lived in urban areas, and this will increase to 65 per cent by 2050: urban people have to be supplied with all of their food.
- People, especially in Asia and in cities, are experiencing increasing **affluence**, meaning they can afford to use more resources like energy.
- In emerging and developing countries, **industrialisation** means a greater demand for energy and water resources.

A key question is whether the world has enough food, energy and water resources to cope with the pressure from more people, urbanisation, industrialisation and affluence. There are two views: pessimistic and optimistic (see Table 7.3).

Now test yourself

What do plants take from soil as they grow, and return to the soil when they die and decay?

TESTED

Exam tip

In the exam you could be asked to make a judgement about which regulating service is the most important.

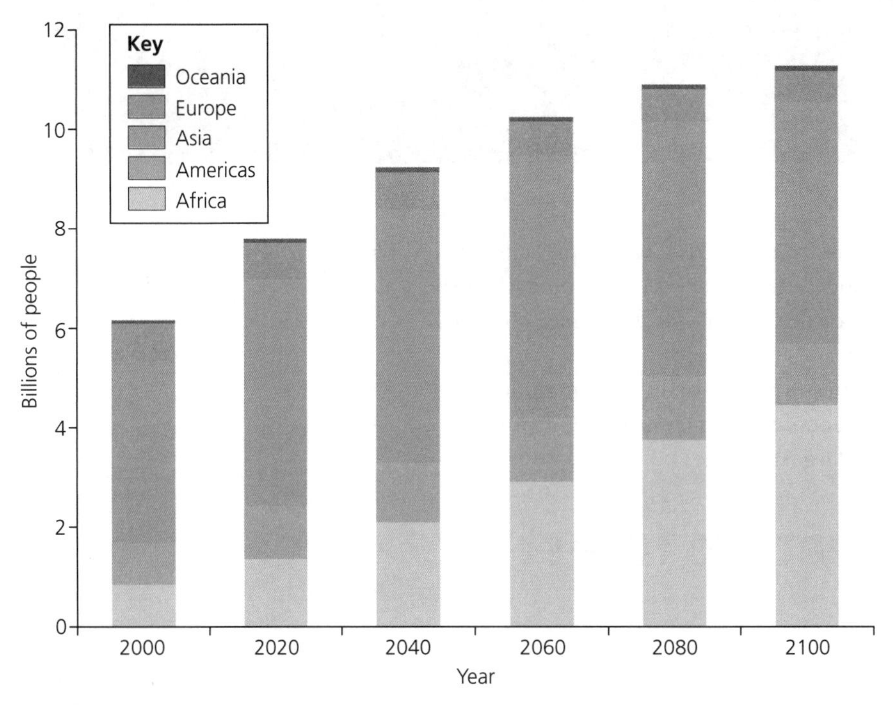

Figure 7.4 Population growth in global regions.

Table 7.3 Theories on the relationships between population and resources

Pessimistic: Thomas Malthus (1798)	Optimistic: Ester Boserup (1965)
Population growth will outstrip food supply, leading to a crisis (famine, conflict, deaths)	Better technology will mean humans can increase food supply to keep pace with population growth
Malthus argued that population would grow geometrically (1, 2, 4, 8, 16, ...) but food production only arithmetically (1, 2, 3, 4, 5, ...)	Boserup argued that 'necessity is the mother of invention'
Population would eventually exceed food supply, leading to a 'crash' in population to re-balance food and people	With more mouths to feed, humans will invent new ways of farming to produce more food: farm machines, fertilisers, irrigation and improved crop strains are examples

It is not possible, yet, to say if Boserup or Malthus is right. The fact that there has been no global crisis so far, as well as advances in farming technology, tends to support Boserup. However, it is yet to be seen whether a world of 10 billion people is possible.

Exam tip

You need to recognise that both Malthus and Boserup could be right, and be prepared to discuss both theories.

Exam practice

1. State two examples of ecosystem goods that indigenous people obtain from the biosphere. [2]
2. Explain the role of the biosphere in regulating the composition of the atmosphere. [4]
3. Explain two reasons why global demand for resources is likely to increase in the future. [4]

ONLINE

Now test yourself

According to Boserup, how much food could the world produce?

TESTED

Topic 8 Forests under threat

Two of the world's three forest biomes are in focus in this topic:
- the tropical rainforest – occurs either side of the equator but within the tropics; occupies thirteen per cent of the Earth's land surface; contains more species than any other biome
- the taiga or boreal forest – occurs only in the northern hemisphere between latitudes 50° and 60°N; occupies seventeen per cent of the Earth's land surface; is the world's largest biome.

What are the threats to forest biomes?

The tropical rainforest

REVISED

The most remarkable feature of the tropical rainforest is its **biodiversity**. There are tens of thousands of different plant and animal species. The key to this biodiversity is the equatorial climate. Throughout the year this climate provides ideal conditions for living organisms.

Biodiversity The number of different plant and animal species in a given area

Adaptations to the environment

All biomes are produced by the interaction of two types of factor:
- Abiotic – the non-living elements, such as climate, water, geology and the mineral matter in soils.
- Biotic – the living organisms, such as plants and animals. Some would also include indigenous or native people.

The tropical rainforest vegetation cover is adapted to make the most of the climate and poor soils. Strong competition between plants, particularly for sunlight, produces a typical vertical structure of four layers (see Figure 8.1). The amazing feature of this structure is that even the 'losers' in the fight for sunlight still manage to survive.

Access to water is another challenge as the thick canopy prevents most rain from directly reaching the forest floor. However, some is transferred indirectly by drip flow from leaf tips in the canopy and stem flow down the tree trunks and lianas.

Now test yourself

TESTED

1 Why is sunlight so important to plant life?
2 Why do trees have buttress roots?

Revision activity

Make brief notes about each of the four layers.

The tropical rainforest offers plenty of food and water for animal life of all sorts throughout the year. Nonetheless, because there are so many animals (mammals, reptiles and insects), there is a great deal of competition for that food. Each of the vertical layers has its own distinctive animal life that is well adjusted to the type of food available in it (see Figure 8.1).

Figure 8.1 The structure of the tropical rainforest.

Animal species have adapted to the challenges of the tropical rainforest environment in a variety of ways, for example by:

- how they move through the forest, for example, monkeys and jaguars
- becoming highly specialised in terms of the food they eat, for example, toucans and parrots
- using various camouflages to avoid being caught and eaten, for example, frogs and lizards
- using bright colours and poisons to warn predators to leave them alone, for example, snakes and spiders.

Now test yourself

TESTED

1 How many layers are there in tropical rainforest, and what are the differences between them?

Functioning

One reason for the high levels of biodiversity is the very high rate of **nutrient cycling** between the three stores: biomass, litter and soil (see Figure 8.2). This is helped by the warm, humid climate.

Nutrient cycling A set of processes whereby organisms extract the chemicals they need for growth from the soil and water, then pass them on through the food chain and eventually back to the soil and water

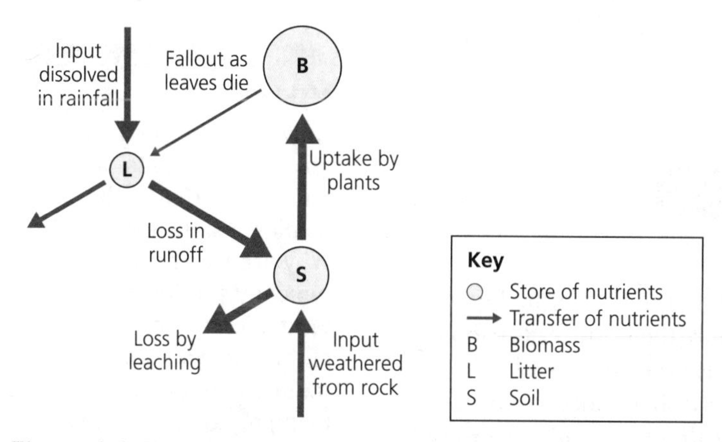

Figure 8.2 The nutrient cycle in the tropical rainforest.

Exam tip

Soils are thin and infertile and therefore contribute little or nothing to the biodiversity.

Now test yourself

2 What are biomass and litter?

TESTED

The high biodiversity is supported by many complex **food webs** which have gradually evolved over hundreds of thousands of years. They ensure that every living creature has a share in the available food supply. But, of course, most of those creatures provide food for the next consumer along a **food chain**.

Food web The many ways in which the plants and animals of a biome or an ecosystem are interconnected. It is made up of many food chains, each one following the path by which a particular animal finds food

Exam practice

1 Describe **two** ways in which **either** plants **or** animals are adapted to conditions in the tropical rainforest. [2]
2 Explain how nutrient cycling helps biodiversity. [4]

ONLINE

The taiga

REVISED

Adaptations to the environment

Like all biomes, the taiga is the outcome of interacting abiotic and biotic factors. Again, some scientists think that the indigenous people are an integral biotic part of the biome. As with the tropical rainforest, the key to understanding the taiga's distinctive features is its climate. It is very different. The main characteristics are:

- short, wet summers
- long, cold, dry winters with temperatures below freezing
- little precipitation
- low sunshine hours
- snow on the ground for months.

Food chain The flow of food (energy) through an ecosystem. Each link in the chain feeds on and obtains energy from the one preceding it. Each link, in turn, is consumed by and provides energy for the following link

Exam tip

Not everyone thinks that people are an integral part of a biome. Rather, most see people as modifiers and exploiters of biomes.

Not many plant species can survive this harsh climate. The notable exceptions are the coniferous trees, such as pine, fir, spruce and hemlock. They dominate. Figure 8.3 shows some of the environmental adaptations that have resulted in the typical dense tree cover of the taiga.

Figure 8.3 Plant adaptations in the taiga.

Given the shortage of edible plants, it is not surprising that animal species are also limited in number. Most animals, notably the reindeer (caribou) and elk (moose), migrate to warmer climates once the cold weather begins, so too do many bird species. Some animals have adapted to the taiga by hibernating when temperatures drop. Others have adapted to the extreme cold temperatures by producing a layer of insulating feathers (for example, owls) or fur (for example, bears) to protect them from the cold.

Now test yourself

1 How does the vertical structure of the taiga differ from that of the tropical rainforest?
2 Why are tree roots typically shallow?
3 Which feature of the taiga climate do you think is most challenging for living organisms and why?

TESTED

Functioning

The taiga differs from the tropical rainforest in three important respects:

- Lower productivity – there are different levels of **net primary productivity (NPP)**. In fact, the figure for the taiga is a little over one-third that of the tropical rainforest. Productivity is greatest when there is plenty of sunlight, high temperatures and precipitation. All three are lacking in the taiga.
- Less nutrient cycling – slower flows and smaller stores in the taiga; largely explained by the much lower temperatures (see Figure 8.4).
- Less biodiversity – this is the outcome of the previous characteristics and the harsh climate allowing plant growth for only four to five months a year. This means little food available for grazing animals.
- Simpler food webs – an outcome of the previous characteristics (see Figure 8.5).

Net primary productivity (NPP) A measure of how much new biomass (plant and animal growth) is added to the biome each year. It is measured in terms of grams per square metre per year

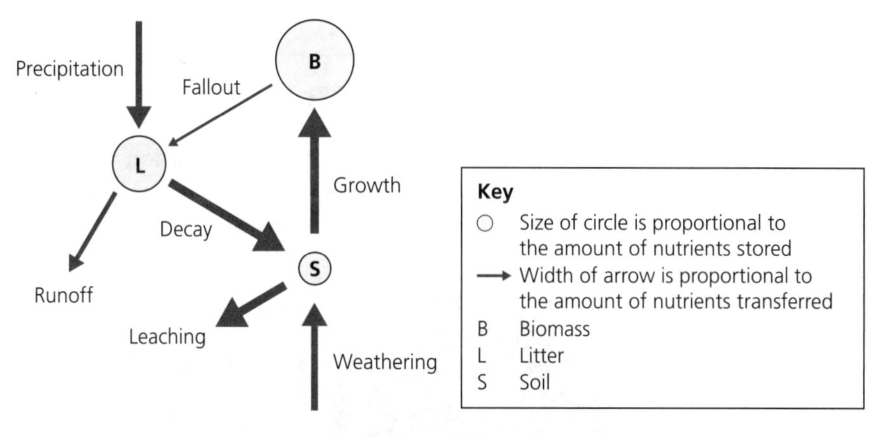

Figure 8.4 The nutrient cycle in the taiga.

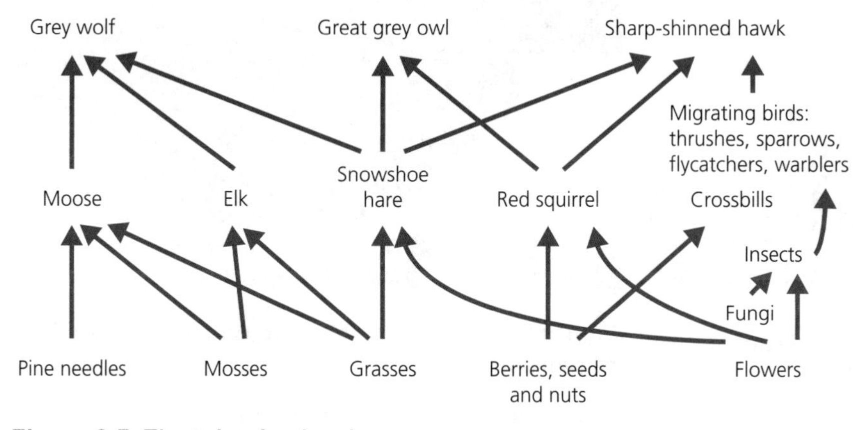

Figure 8.5 The taiga food web.

Revision activity

Compare Figures 8.2 and 8.4 and the nutrient cycles they show. Make notes about the differences in a) stores and b) flows.

Exam practice

1 Identify **two** ways in which coniferous trees cope with heavy snowfall. [2]

2 Suggest reasons why the taiga food webs are simpler than those of the tropical rainforest. [4]

ONLINE

Now test yourself

1 What are the factors controlling the rate of nutrient cycling?

2 Using Figure 8.5, give an example of a) a primary consumer and b) a secondary consumer.

TESTED

The tropical rainforest under threat

REVISED

The biggest direct threat to all forests is **deforestation**. Between 1990 and 2015, the world lost 129 million hectares of forest. That is an area the size of South Africa. Most of this deforestation took place not in the tropical rainforest, but in the taiga.

Deforestation
The deliberate clearance of forested land for conversion to another land use

Causes of deforestation

The two main causes of deforestation are:

- economic development
- population growth.

Together these are increasing the demand for resources that happen to occur in the tropical rainforest. These include timber, oil, gas, iron ore and gold. But ironically, it is none of these that is driving much of the deforestation.

Case study: deforestation in the Brazilian Amazon

Brazil is a vast country and is the 'owner' of by far the largest share of the world's tropical rainforest (25 per cent). The annual losses to deforestation here have been huge (see Figure 8.6). But on the positive side it can be seen that the amount today is much less than it was in 2004.

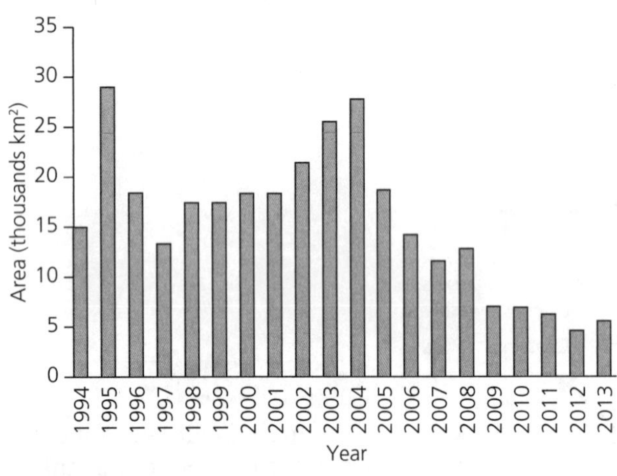

Figure 8.6 The annual rate of deforestation in the Brazilian Amazon from 1994 to 2013.

Figure 8.7 shows that the most valuable resource of the tropical rainforest in Brazil is in fact the land underneath it. Cleared land is in high demand to create grazing for livestock and, to a lesser extent, arable land for the growing of crops (see Figure 8.7). It is the global need for more food to feed a rapidly growing population that is the overwhelming cause of deforestation in this vast tract of the tropical rainforest.

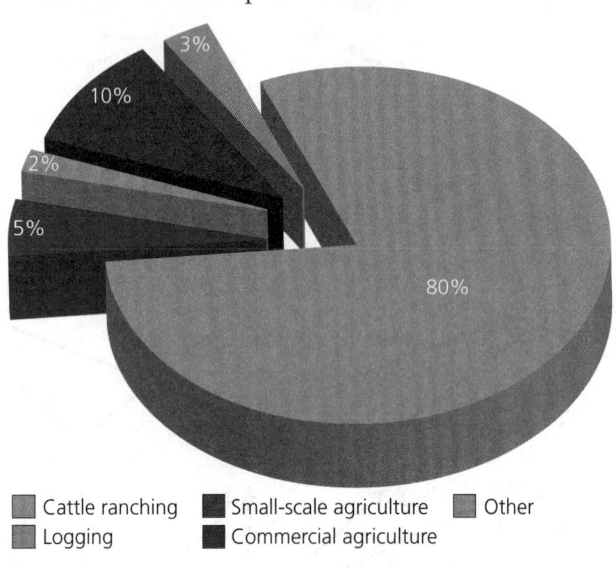

Figure 8.7 Causes of deforestation in the Brazilian rainforest.

Another cause not specified in Figure 8.7 is the drowning of forested valleys to create giant reservoirs for the generation of hydroelectric power. This land conversion is included in the 'other' category, along with the building of new settlements and roads.

Valuable though hardwoods are, commercial logging is not the main cause of deforestation. Figure 8.7 puts that cause into perspective. But, of course, the timber that is felled to make way for farming or reservoirs is sold and exported.

The problem with land cleared for pasture and crops is that the soil quickly declines in fertility. This means that after a few years, the farmers have to cut down more rainforest for a new supply of farmland.

Climate change

One of the global impacts of the clearance of the tropical rainforest is on climate change. If undisturbed, the trees absorb huge amounts of carbon dioxide in the atmosphere. This stops, of course, as soon as the trees are felled. More carbon dioxide in the atmosphere means more global warming. But looking at the relationship in the opposite direction – in what ways does global warming threaten the rainforest?

It is expected that global warming will increase the incidence of drought in rainforest areas. The Amazon rainforest suffered three severe droughts in 2005, 2010 and 2014. During them, the rainforest changed from absorbing carbon dioxide to emitting it. Forest fires broke out in the dry conditions. The burning of trees and forest litter released large amounts of carbon dioxide.

Revision activity

Note that among the crops being grown on cleared land in Brazil are biofuels, such as sugar cane and soybean (see Topic 9, page 133).

Now test yourself

TESTED

1 Why do you think there is much more concern about the loss of tropical rainforest than the loss of the taiga forest?
2 How does deforestation cause soil erosion?
3 Which activities are likely to cause the most serious river pollution?

Other repercussions of the droughts include:

- nutrient cycles threatened by decomposers dying as litter stock drying out (see Figure 8.2, page 121)
- food supply and food webs disturbed by leaves in the canopy dying
- less evaporation and transpiration from dying trees, so less cloud and rain.

Scientists now recognise that climate change will:

- do permanent damage to the rainforest
- cause the rainforest distribution to contract towards the equator and be replaced by tropical grasslands
- cause more fires
- switch the role of rainforests from carbon sink to carbon source
- endanger many animal and plant species and so reduce biodiversity.

In short, climate change and deforestation threaten to accelerate the rate of global warming.

Exam practice

1 Assess the causes of deforestation in the tropical rainforest. [8]
2 Explain how the clearance of tropical rainforest is leading to global warming. [4]

ONLINE

The taiga under threat

REVISED

Direct and indirect threats

Unlike the situation in the tropical rainforest, the main cause of deforestation in the taiga is logging to provide:

- wood pulp for making paper
- building timber
- wood chips for making chipboard, fibreboard and biofuel.

The scale of deforestation caused by the logging industry is huge. Between 2000 and 2013, Canada and Russia alone accounted for just over 40 per cent of all the global deforestation.

So why is there less concern about the deforestation here than in the tropical rainforest? The answer lies in the following:

- The taiga biome is the largest in the world; only eight per cent of it has been cleared so far.
- The taiga is situated in some of the remotest places on Earth and so is 'out of sight'.
- The taiga has a much lower level of biodiversity. In other words, more species are at risk in the tropical rainforest. The media have done much to focus public attention on endangered species there, such as jaguars, monkeys and parrots.

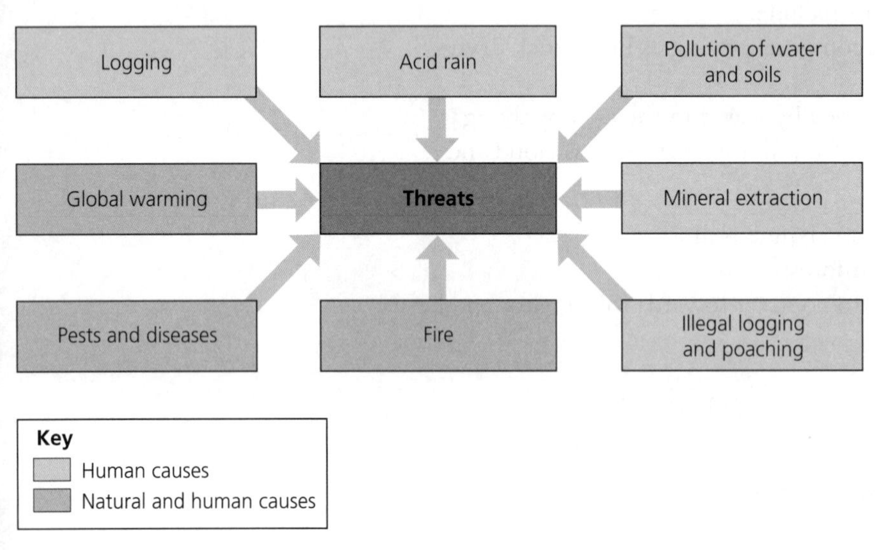

Figure 8.8 Threats to the taiga.

The taiga is also subject to indirect threats (see Figure 8.8). For example:

- the pollution of rivers and lakes by the chemicals used in the production of wood pulp
- the **strip mining** of minerals, such as tar sands, coal, nickel, lead and cadmium
- oil and gas extraction with drilling rigs, pipelines and oil-spill accidents (see Topic 9, page 140)
- hydroelectric power developments.

Strip mining Used to extract minerals located near the Earth's surface. The overlying rock and soil are removed to expose the mineral deposit which can then be extracted easily and cheaply. It creates great scars on the landscape

Now test yourself

TESTED

How do hydroelectric power developments threaten the taiga?

A loss of biodiversity

Biodiversity in the taiga is much lower than in the tropical rainforest. Even so, there are threats to it that are not all directly related to human use of the forest and its resources. These threats are illustrated in the following case study of the Russian taiga (see Figure 8.8).

Case study: threats to biodiversity in the Russian taiga

Deforestation in Russia is occurring at a rate of roughly 20,000 km^2 per year. Logging and the clearing of land for agriculture are the two main reasons for it. But there are other threats:

- Illegal logging and poaching – the high prices paid for forest products have encouraged a lot of illegal logging. This means that much more of the forest is being cut down than allowed by the annual logging quotas. Also poaching both by hungry local people and professional hunters is upsetting food webs.
- Fires might be thought of a threat but in fact they are part of the natural regeneration process. The taiga is surprisingly prone to wildfires for a number of natural reasons:
  - they are often started by lightning strikes during the short summer
  - the thick carpet of pine needles easily catches light
  - the trees contain a resin which burns very easily.

  Fires in the Russian taiga are becoming more frequent and increasing in scale. This is being blamed on global warming with its hotter and drier summers. But people are also to blame. For example, hunters leave their campfires unattended and energy companies allow oil spills and gas leaks to ignite (see Topic 9, page 140). No matter what the cause, these large-scale fires release huge amounts of greenhouse gases and are therefore contributing to global warming.
- Acid rain is a human-made threat. It is a result of fossil fuels being burnt by industries located in the taiga, such papermaking and the smelting of metallic ores. Once acid rain gets into the soils, lakes and ponds, it can stress and kill trees over large areas.
- Pests and diseases are a normal part of the taiga biome, but they are on the increase. Again climate change is being blamed. Because winters are becoming slightly less cold, many more pests and diseases are now surviving from one year to the next. Some pests and diseases are accidentally spread by people. They, along with acid rain, are reducing biodiversity and creating a more open landscape.
- An added threat in the Russian taiga is contamination by nuclear waste released during nuclear weapon tests in the 1950s and 1960s.

All these threats are reducing the forest and its biodiversity. In doing so, they are changing its character and endangering rare animal species such as the Siberian crane, the red wolf and the Amur tiger.

Now test yourself

TESTED

1. Which threats to biodiversity are mainly the outcome of human activity?
2. How do dying trees impact on the food web?

Exam tip

With the exception of nuclear waste, the threats in Russia are very much the same in northern Europe, Alaska and Canada, but less severe and out of control.

Exam practice

1. Explain why the remoteness of much of the taiga is a disadvantage. [4]
2. State three ways in which global warming is threatening the taiga. [3]

ONLINE

Conservation and sustainable management of the tropical rainforest

REVISED

Global actions

There are two international agreements designed to conserve forests and to protect the species within them:

- Convention on International Trade in Endangered Species (CITES) – this bans cross-border trading in some 34,000 endangered plant and animal species. By stopping the buying and selling of the most threatened species, the hope is to stop illegal hunting and collecting. This approach to **conservation** has its advantages and disadvantages (see Table 8.1).
- Reducing Emissions from Deforestation and Forest Degradation (REDD) – this United Nations project adopts a rather different approach. It is aimed at stopping the clearance and degradation of forests, but on the grounds of checking global warming. Governments and transnational corporations (TNCs) fund forest conservation projects in developing countries. By doing so, they can offset their own carbon emissions. The outcome is that developed countries pay for putting the brake on deforestation.

Conservation The protection and possible improvement of biomes, landscapes or environments for future benefit

Table 8.1 The advantages and disadvantages of a CITES approach

Advantages	Disadvantages
Has the backing of 180 countries	Protection of species does not halt deforestation
Protection applies to species in all biomes, not just tropical rainforest	More concern today about reducing the causes of global warming
Works well with limelight species, such as tigers and chimpanzees	Governments have to do the policing. Not all developing governments have the resources
Successful in outlawing the ivory trade and trade in rare parrots	Species not listed until they are on the brink of extinction – often too late

The problem with both CITES and REDD is that they are difficult to enforce. Much illegal activity can go unnoticed in remote and less accessible locations. It is this variation in the enforcement of international agreements that helps to explain why deforestation rates are rising in some areas (poor policing) and falling in others (effective policing).

Now test yourself

TESTED

Which of CITES or REDD do you think is more successful in protecting biomes? Give your reasons.

Sustainable forest management

CITES and REDD may be described as top-down approaches to conservation. Much can be done by working in the other direction. Bottom-up, grass-roots approaches are essentially local.

In looking at what can be done at a grass-roots level, it is important to distinguish between two different situations:

- areas being logged
- areas untouched by logging.

For the first, there are at least three **sustainable forest management** strategies:

- Selective logging – felling trees only when they are fully grown and letting younger trees mature.
- Agroforestry – allowing crops to be grown in carefully controlled cleared areas along with trees that may be harvested for fuelwood and building timber.
- Replanting (reforestation) – re-create the forest cover by collecting seeds from the remaining primary forest, growing the seeds into saplings in nurseries and then replanting the saplings in deforested areas.

All three actions offer local people livelihoods.

Exam tip

Between the international and local scales, governments can contribute by designating national parks, nature reserves and various types of protected area.

Sustainable forest management Organising the use of forest resources in such a way that they will be available for the benefit of future generations

Ecotourism An environmentally friendly and alternative form of tourism that seeks to minimise the ecological impacts of tourists and the consumption of non-renewable resources

Now test yourself

TESTED

1 Which of these three sustainable forest management activities offers the most secure livelihood? Give your reasons.

For those areas untouched by logging, **ecotourism** presents a form of sustainable action. Scenery, wildlife, remoteness and culture are the main attractions. It aims to educate visitors and increase their understanding of nature and local cultures. Since the spirit of ecotourism is that it is controlled by local people, employs local people and uses local produce, and its profits stay in the local area, it offers a viable livelihood that does not threaten the rainforest.

Now test yourself

TESTED

2 Suggest two possible disadvantages of ecotourism.

Exam tip

Remember that the tropical rainforest still contains native tribes that have been living in harmony with the biome for thousands of years. Their survival must be a top priority in a sustainable management programme.

Exam practice

1 Explain the difference between the CITES and REDD approaches to the conservation of tropical rainforest. [4]

2 Assess the relative merits of three different forms of sustainable forest management. [8]

ONLINE

Protection of the taiga wilderness

REVISED

Creating and maintaining wilderness

The taiga's vast scale and remoteness can make it easy for us to think that the forest can look after itself. Because of its lack of human settlement, most of it qualifies as **wilderness**. However, Figure 8.8 (page 126) shows that the taiga is being threatened in a number of different ways. It can no longer rely on isolation and extent to give it the protection it deserves.

Wilderness Areas that are isolated and inaccessible and where there is little or no human activity or settlement

Ramsar site A wetland site protected by the Ramsar Convention, an agreement signed in 1971

Two questions arise from this situation:

- Does wilderness have a value?
- How best two protect it?

The protection of wilderness in the taiga may be justified because of the need:

- to conserve biodiversity and ensure that endangered species survive
- to provide opportunities for people to enjoy the wilderness experience – a temporary escape from the pressures of modern living
- to maintain the tree cover as much as possible as a check against the build-up of too much carbon dioxide in the atmosphere.

Now test yourself

Why is the last of these three needs important?

Wilderness is best protected by the setting up of national parks or biosphere reserves that cherry-pick some of the best stretches of untouched biome. Canada, Russia and the USA have all taken this course of action.

Generally speaking, a national park:

- covers an area of more than 1000 hectares
- has legal protection and financial support
- is open to the public for recreation and leisure, but in a controlled way.

The protection is not just about wildlife and their habitats, but should also apply, as in the tropical rainforest, to the indigenous or native tribes, such as the Sami in northern Europe and the Inuits in North America. In Canada, such people are referred as First Nations.

The protection of wilderness can be strengthened if the proposed area can justify some form of international recognition, for example as a **Ramsar site** or World Heritage site. A less satisfactory protection lies in being declared an 'area of sustainable forestry'.

Revision activity

Make notes about the types of location that qualify for Ramsar site or World Heritage site status.

Conflicting views

Figure 8.9 looks at the relationships between seven competing stakeholders in the taiga. Broadly speaking, they fall into two conflicting groups:

- those keen to exploit the forest resources for profit
- those keen to conserve and protect this huge biome and its native people.

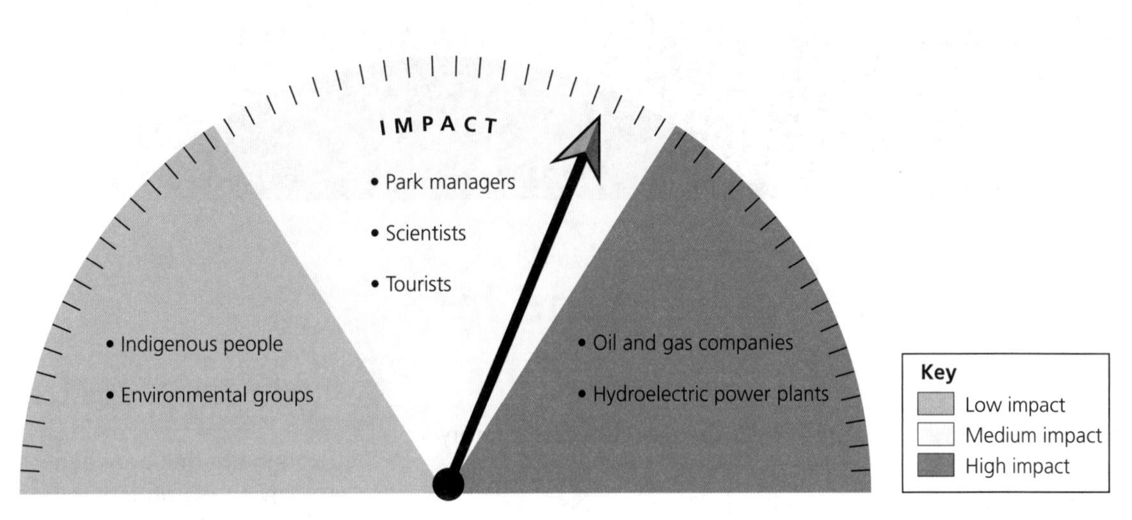

Figure 8.9 Competing stakeholders in the taiga.

The only way to resolve the conflicts shown in Figure 8.9 is to find some sort of compromise or middle path between the two main groups. It is the duty of governments to make sure this happens.

Most in the conservation group would be content with:

- the creation of a network of national parks or biosphere reserves, each with its buffer and transition zones
- strict policing in those parks and reserves to ensure that protection is real and effective
- everything being done to make the working of the taiga's resources as sustainable as possible.

So what might the last of these conditions involve? The possible actions are much the same as in the tropical rainforest:

- abandoning the practice of clear-felling undertaken by most lumber companies; selective logging should mean less environmental damage in the form of soil erosion and landslides
- reforestation – planting trees in felled areas
- minimising the fire risks associated with all activities
- preventing the pollution of air and water.

Revision activity

Figure 8.9 does not include the logging industry.
How would you score its relationships with the other seven stakeholders?

Now test yourself

TESTED

1 How does selective logging reduce the likelihood of soil erosion?

None of these objectives is impossible if all the economic activities recognise that they have a responsibility to minimise their footprints. Becoming more sustainable may mean extra costs, but is the long-term survival of the taiga not worth it?

Exam tip

Remember it is not the wish of conservationists to protect the whole of the taiga. So why should some of its resources not be used?

Exam practice

1 Explain the value of the taiga wilderness. [4]
2 Assess the chances of reaching a compromise about the taiga's future. [8]

ONLINE

Now test yourself

2 How might an oil company justify being allowed to operate in the taiga?

TESTED

Topic 9 Consuming energy resources

How can the growing demand for energy be met?

The global demand for energy has tripled over the past 50 years, mainly for three reasons:

- global population growth
- rising levels of development and living standards
- modern technology which consumes energy but gives more people access to various forms of energy.

Now test yourself

1 Explain the link between energy consumption and living standards.

TESTED

Classifying energy resources

REVISED

Energy is needed for a whole range of important activities. Figure 9.1a shows the major energy users. Supply of that energy comes from a range of different primary sources (see Figure 9.1b).

Now test yourself

2 Why is access to cheap, reliable energy important to a country and its people?

TESTED

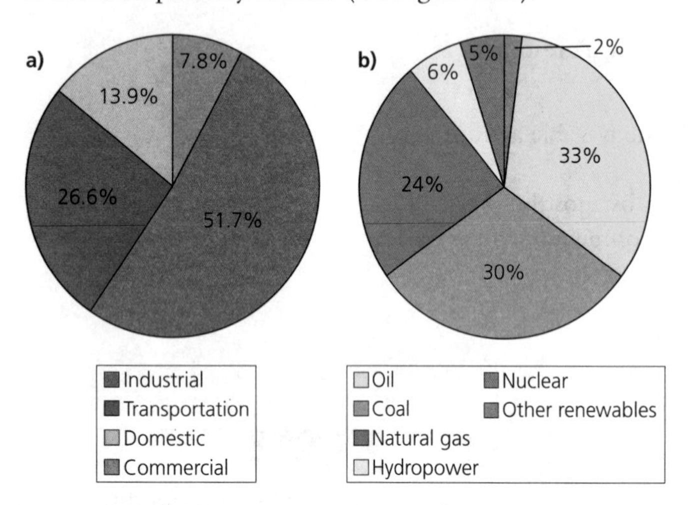

Figure 9.1 a) Global energy users; b) global energy supply.

As countries become more developed, they use more energy. Access to cheap and reliable energy resources is important to the 'health' of a country's economy and to the well-being of its people.

Non-renewable Energy produced from sources that cannot be replaced once they are used

Renewable A natural energy source which is not diminished when it is used; it recurs and cannot be exhausted

Recyclable Sources of energy that are renewable but only as a result of human actions, such as the planting of trees for fuelwood and the growing of biofuel crops. Also includes nuclear energy because the nuclear fuel can be recycled

Three types of energy resource

Today's supply of energy comes from three different types of source:

- **non-renewable** – for example, coal, oil and gas (these are the so-called fossil fuels)
- **renewable** – for example, water, wind, solar and tidal power (energy flows)
- **recyclable** – for example, nuclear energy, biomass and biofuels.

Exam tip

Be sure that you have a clear understanding of the term 'recyclable energy'.

Now test yourself

TESTED

3 What is meant by the term 'energy resource'?

Environmental impacts

Fossil fuels are renowned for their **carbon emissions** into the atmosphere. But the use of all energy resources has environmental costs of some sort or another. The natural environment is changed in some way, usually negatively, for example by:

- the working of coal – the mines and the huge open-cast workings, as in Australia
- the drilling for oil and gas – the rigs and the added risk of oil spills or gas leaks, as in Alaska
- the clearance of tropical rainforest to grow biofuels, as in the Amazon basin
- the scarring of scenery by wind and solar farms, as in the Scottish Highlands
- the flooding of valleys by dams to provide hydropower, as in the Three Gorges project, China.

Carbon emissions Strictly speaking, these are amounts of carbon dioxide released into the air when fossil fuels are burnt. Carbon dioxide is a greenhouse gas

Revision activity

Make sure you know two energy resources in each of the three categories: non-renewables, renewables and recyclables.

Now test yourself

TESTED

1 What is the problem with carbon emissions?

Most of the energy resources in all three categories are converted into electricity. The attraction of electricity is that it is a clean form of energy and easily distributed. However, converting primary sources of energy into electricity does have environmental impacts (see Figure 9.2).

	Biomass	Coal	Nuclear	Natural gas	Solar	Wind
Climate change impact	Moderate	High	Low	High	Low	Low
Air pollution impact	Moderate	High	Low	Moderate	Low	Low
Land impact	Moderate	High	High	Moderate	Moderate	Moderate
Water impact	Moderate	High	High	High	Low	Low
Other impacts	Moderate	Moderate	High	Moderate	Low	Moderate

Figure 9.2 Comparing the environmental impacts of electricity generation.

Now test yourself

TESTED

2 Which energy resource has the greatest environmental impacts when converted into electricity?

Exam tip

It is important to remember that all forms of energy supply actually use energy during:

- the construction of energy plants (for example, transporting wind turbines to wind farms)
- the day-to-day running of energy plants (for example, power stations)
- the distribution of energy (delivering coal, oil and gas).

Exam practice

1 Identify four domestic uses of energy. [4]

2 Assess the costs and benefits of producing energy from fossil fuels. [8]

ONLINE

Access to energy resources

REVISED

It is the aim of most countries to use, as much as possible, any energy resources located within their borders. However, whether or not those resources are exploited is affected by many factors.

Factors affecting access to energy

The factors fall into two broad categories: physical and human (see Figure 9.3).

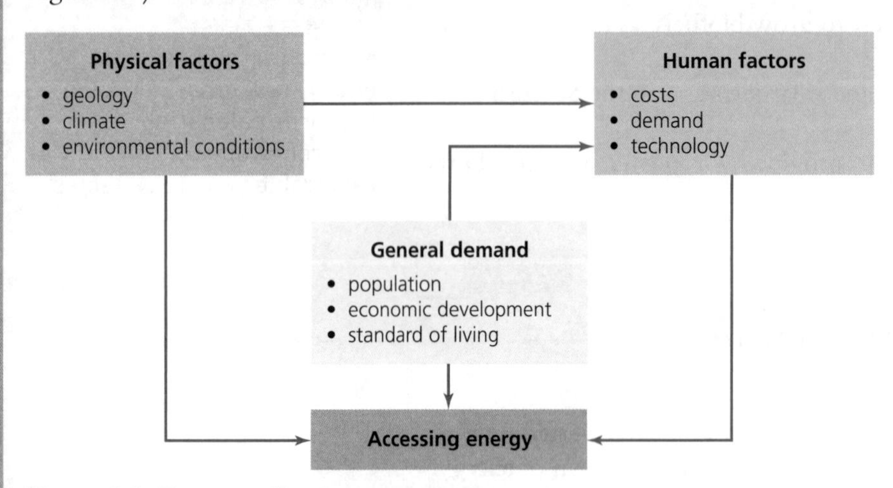

Figure 9.3 Factors affecting access to energy.

Physical factors include:

- Geology – this determines, for example, whether or not a country contains deposits of fossil fuels. Whether or not those deposits are actually used depends on other factors such as:
  - can they be easily worked?
  - how extensive are they – can they be worked profitably?
  - how large are the **reserves**?
  - is there capital and technology available to exploit those reserves?
- Climate – is the climate able to provide reliable renewable sources of energy, such as water (rainfall), sunshine hours and windy days?
- Environmental conditions – access to energy sources can be made difficult by a harsh climate and a mountainous terrain.

Human factors include:

- Costs – basically the costs of exploiting an energy resource and delivering that energy to its users. The cheaper (more accessible) energy resources are likely to be exploited first and most. As shown in Figure 9.3, costs are influenced by both physical and human factors.
- Level of demand – a high demand for energy usually means good profits for the energy company. Some of those profits can then be reinvested in accessing new or more difficult energy sources.
- Technology – this can bring four benefits:
  - an ability to tap less accessible energy sources
  - a lowering of costs due to the efficiency of modern equipment and know-how
  - the discovery of new sources of energy
  - the invention of new ways of generating energy.

Reserves Part of a natural resource considered to be exploitable given present economic conditions and available technology

Now test yourself

Which renewable source of energy is affected by geology?

TESTED

Revision activity

Make notes about one location where environmental conditions are making it difficult to access energy resources.

Exam tip

The costs of accessing energy are most important. Be sure you understand what affects those costs.

Global variations in energy use

Figure 9.4 shows that the use of energy per capita varies greatly in different parts of the world.

Two possible causes of this global pattern have already been identified:

- levels of development and standard of living
- available technology.

Basically, the more developed a country, the greater its use of energy. Much energy is going to be used by the sectors of its economy:

- in agriculture – mechanisation
- in manufacturing – processing commodities into products
- in the tertiary sector – transport, heating and lighting of shops, offices and homes.

A developing country is likely to use less energy, but it is also likely to rely more on traditional fuel resources, such as fuelwood.

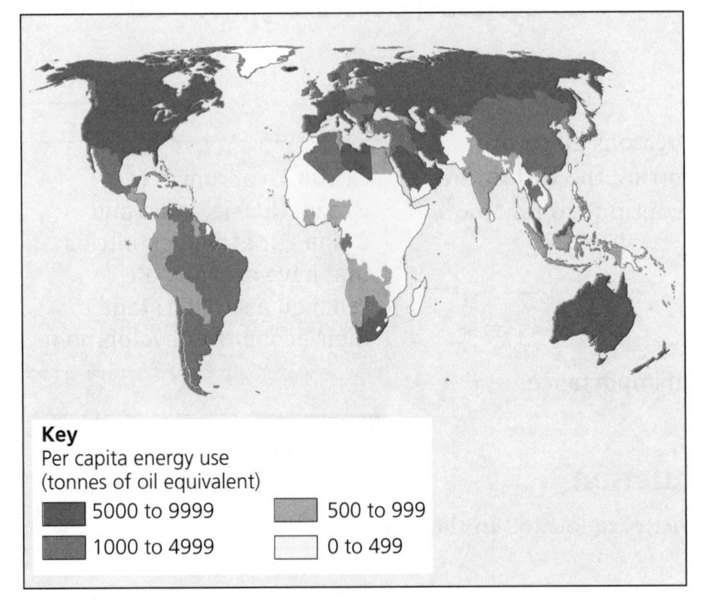

Figure 9.4 The global distribution of energy use.

There is, however, another very significant factor – climate. Where temperatures are low and daylight hours short, energy is needed for heating and lighting. Where temperatures are high, energy may be needed instead for air conditioning. Clearly, countries in the middle latitudes have an advantage here.

Now test yourself

TESTED

2 Figure 9.4 shows a low level of energy use over much of Africa. What are the reasons for this?

Exam practice

1 Identify three elements of the UK climate that could be sources of renewable energy. [3]

2 Study Figure 9.4. Suggest reasons for the high level of per capita energy use in Australia. [4]

ONLINE

Now test yourself

1 Why is population size not considered to be one of the factors?

TESTED

Revision activity

Study Figure 9.4. Make notes of the main areas where per capita energy use is highest.

The rising global demand for oil

REVISED

The history of energy since the beginning of the Industrial Revolution in the mid-eighteenth century has centred, in turn, on three fossil fuels:

- coal – up to the mid-twentieth century
- oil – in the second half of the twentieth century
- gas – this is beginning to challenge oil as the top fossil fuel in the early twenty-first century (see Figure 9.5).

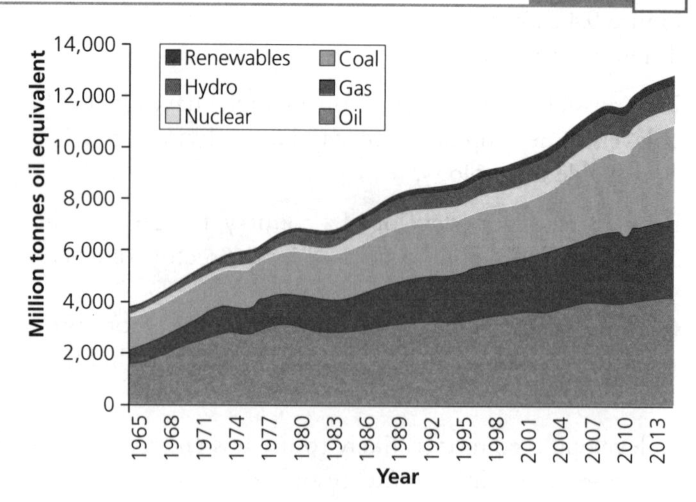

Figure 9.5 Sources of global energy, 1965–2013.

The use of oil continues to rise because it has a number of advantages as a source of energy:

- it burns more cleanly than coal – lower carbon emissions
- it is fairly transportable – by pipeline, road, rail and ship
- it has a number of possible uses – transport, domestic, industrial and generating electricity.

The rise is also explained in part by the same three reasons given on page 135. However, the four leading emerging countries, the **BRICs**, are divided when it comes to energy. China and India continue to rely more heavily on coal, Brazil and Russia on oil.

BRICs An acronym for Brazil, Russia, India and China: emerging countries that have recently all reached a similar stage in their economic development

Now test yourself

TESTED

1 Name the three main energy sources in order of importance.

Exam tip

Be sure you know the main uses made of oil.

Distribution of oil reserves and production

Table 9.1 shows the top ten oil producers. Half of them are located in the Middle East – the oil capital of the world.

When it comes to oil reserves, the pattern is much the same, except that a few 'new' countries appear in the top ten rankings, for example Venezuela, Libya and Nigeria, displacing the USA, China and Brazil.

Table 9.1 The top ten oil producers and oil reserve holders, 2016

Country	Production (thousands of barrels per day)	Country	Proven reserves (billions of barrels)
Saudi Arabia	10,625	Venezuela	294.4
Russia	10,254	Saudi Arabia	268.3
USA	8,744	Canada	171.0
Iraq	4,836	Iran	157.8
China	3,938	Iraq	144.2
Iran	3,920	Kuwait	104,2
Canada	3,652	Russia	103.2
UAE	3,188	UAE	97.8
Kuwait	3,000	Libya	48.4
Brazil	2,624	Nigeria	37.1

Now test yourself

2 Suggest reasons why Venezuela, Libya and Nigeria are not among the top ten oil producers.
3 Using the data in Table 9.1, calculate the amount of oil production that comes from the Middle East.

TESTED

Oil supply and prices

Figure 9.6 shows how the price of crude oil has changed since the year 2000. The variations are largely explained by the laws of **demand** and **supply**. When supply exceeds demand, the price of oil falls, as between 2014 and 2016. When demand exceeds supply, prices rise, as between 2009 and 2011. All of these ups and downs coincide with periods of recession (falling prices) and boom (rising prices).

These changing relationships between demand and supply are affected by the fact that oil is moved between the few producer countries and the many consumer countries. The fortunes of this international trade rely on:

- the diplomatic relations between the trading partners; any upset of these relations can disrupt the trade
- political stability; conflicts in the Middle East over the past 25 years have seriously disrupted the oil trade and affected crude oil prices.

Demand The wish to acquire a specific good or service at a particular price

Supply The amount of a specific good or service that is available to consumers

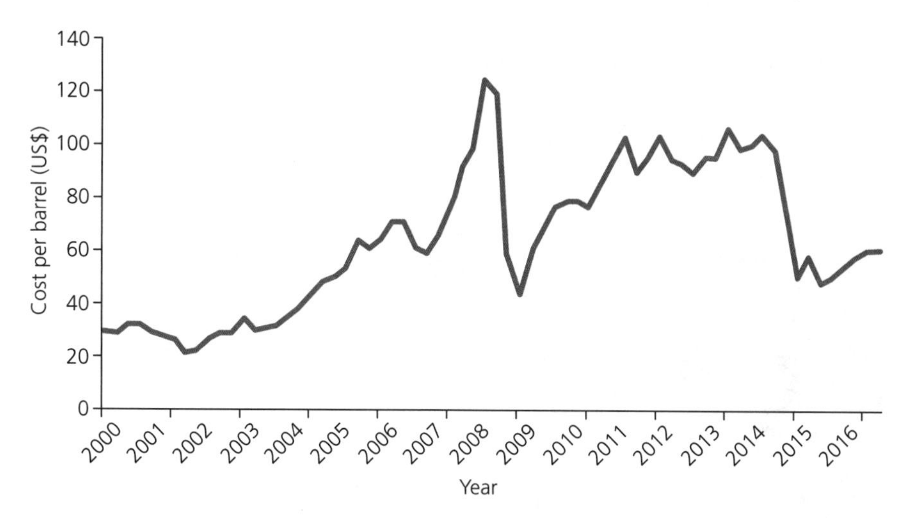

Figure 9.6 Crude oil prices from 2000 to 2016.

Now test yourself

TESTED

What happens when a country's oil production is interrupted by conflict?

Exam tip

Remember what Figure 9.6 shows: the price of oil has ranged roughly from $20 to $120 per barrel, that is a huge 500 per cent increase since 2000.

Exam practice

1 Describe the advantages and disadvantages of relying on oil as an energy source. [4]

2 Suggest reasons why the global production of oil varies from year to year. [4]

ONLINE

Continuing reliance on fossil fuels

REVISED

Despite growing global concern about the use of fossil fuels, energy companies are still looking for:

- new reserves of **conventional** oil and gas
- new **unconventional** sources of oil and gas.

In both cases, the aim is to find resources that are accessible and therefore allow companies to continue to supply both fuels at competitive prices.

Conventional In the case fossil fuels this means the usual working of coal by mining and of oil and gas by drilling in the usual locations

Unconventional In the case fossil fuels this means doing things differently, in terms of both locations and methods of extraction

New conventional sources of oil and gas

The fact that oil and gas are non-renewable resources means that energy companies are always searching for new reserves of these conventional sources. As one oilfield becomes exhausted, another one has to come on stream.

The search is focusing on three types of location (see Figure 9.7):

- in the Arctic regions of North America and Eurasia – new oil and gas fields are beginning to come on stream in northern Siberia and Alaska
- in inland arid areas such as central Siberia
- in coastal areas, such as around the Gulf of Mexico and along the Skeleton coast of Angola and Namibia (Africa).

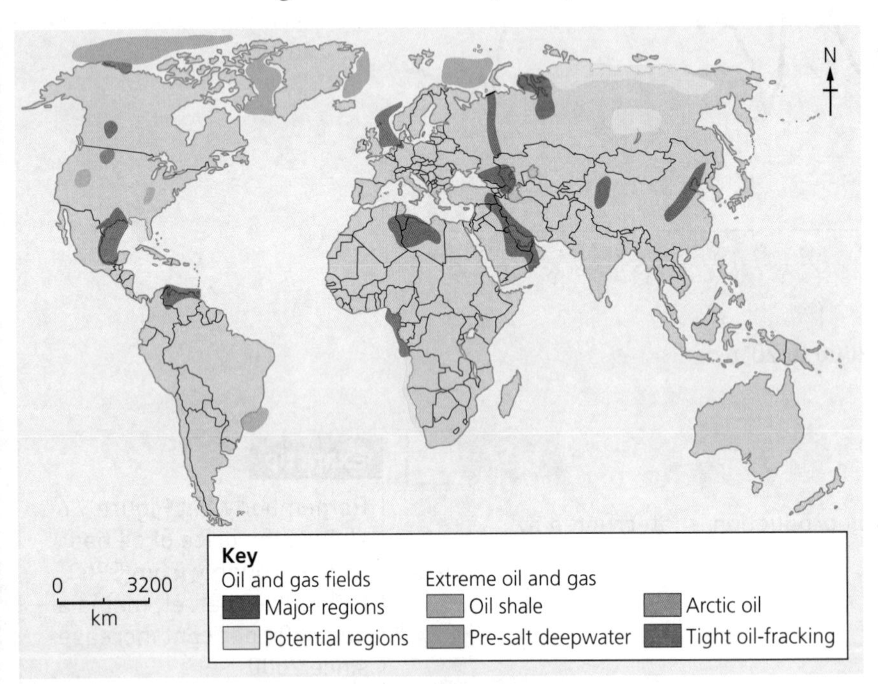

Figure 9.7 Oil and gas fields: present and future.

Because of their locations, these newly discovered oil and gas fields present some major challenges. They include:

- working in extremely cold and hostile environments
- transporting oil and gas over long distances that isolate the new fields from the major markets
- disturbing and damaging sensitive environments (see the case studies that follow).

These challenges might be looked on as some of the costs of developing these new sources. These need to be set against the undoubted benefit of prolonging the global supply of oil and gas.

Exam tip

You might question whether we should be searching for new reserves of oil and gas. Should we not be searching instead for new sources of renewable energy?

Now test yourself

Name a) two major oil and gas regions, and b) two potential oil and gas regions.

TESTED

New unconventional sources of oil and gas

The four unconventional sources included here are shown in Figure 9.7 under the heading of 'Extreme oil and gas'.

The four unconventional sources of oil and gas are shown in Table 9.2. Exploitation is still in its early days. However, there is already a great deal of public concern, particularly about their possible impact on the environment. For example:

- the large-scale and disfiguring opencast mining that is associated with tar sands and shale gas
- the pollution of groundwater by the fracking process used to obtain shale gas
- the exposure of deepwater oil rigs to extremely stormy seas means a high risk of oil spills and pollution
- the disturbance or destruction of sensitive ecosystems, such as the temperate grasslands and taiga of North America
- the opening up of remote wilderness areas which conservationists say should be left to nature.

Table 9.2 Unconventional sources of fossil fuels

Resource	Nature	Extraction methods
Tar sands	A mixture of clay, sand, water and bitumen (a heavy, viscous oil)	Needs to be mined then injected with steam to make the tar less viscous so that it can be pumped out
Oil shale	Oil-bearing rocks that are permeable enough to allow the oil to be pumped out directly	Can be mined or the shale is ignited so that the light oil fractions can be pumped out
Shale gas	Natural gas that is trapped in fine-grained sedimentary rocks	Needs to be fracked: water and chemicals are pumped in to force out the gas
Deepwater oil	Oil and gas that is found a long way offshore at extreme depths in the sea	Drilling from rigs; already being done in the Gulf of Mexico and off Brazil

The problem here is uncertainty. Only time will tell whether or not the threats are real or imagined. There are as yet few hard facts, but much guesswork.

Exam tip

Try to be fair in your evaluation of these new sources. Be careful about jumping too quickly on the 'anti' bandwagon. Weigh up the available evidence.

Now test yourself TESTED

Which of the possible environmental impacts of these four unconventional sources of energy do you think is the most serious? Give your reasons.

Revision activity

Create a spider diagram containing notes about advantages and disadvantages of each of these four alternative sources of fossil fuels.

The two case studies that follow illustrate the tensions that exist between the use of energy resources and forest biomes and use of the energy resources that occur within them.

Case study: oil and the Amazon tropical rainforest

The development of oil reserves by Ecuador and Peru in the western Amazon shows how damaging the industry can be. The main environmental impacts are:

- cutting roads through the forest to the oil rigs
- the roads and oil rigs attract migrants and their settlements lead to further deforestation
- the gases that are the by-product of oil drilling are burned off and this pollutes the local air and increases the risk of forest fires
- spills from burst pipelines do much ecological damage
- toxic drilling chemicals are dumped and cause much pollution of soils and rivers.

Brazil owns most of the Amazon rainforest, but as yet there has been little oil exploitation. The main reason is that it has more accessible offshore oil reserves. But energy is creating a rather different kind of threat. This is the clearance of rainforest to grow biofuels.

Case study: oil and the taiga

Russia is estimated to have twenty per cent of the world's oil and gas. Most of its reserves are located in the taiga. Exploitation has meant cutting great swathes through the forest to make way for huge and extremely long pipelines. These carry the oil and gas from the remote 'fields' either to ports or directly to the major markets as far afield as western Europe.

Leaks from these pipelines are not uncommon, nor are spills at the drilling sites. These accidents are particularly damaging because drainage in the taiga is very poor. This means that the oil does not get washed away; it is slow to decompose. Instead it seeps down into the soil and is taken up by the shallow roots of the coniferous trees, often killing them.

Although Canada and the USA (in Alaska) have a better record of fewer oil spills and pipeline leaks, there is an added threat in Canada. This is the large-scale clearance of forest to make way for the opencast working of tar sands in the Athabasca area of Alberta.

Exam practice

1. Describe the costs and benefits of developing new oil and gas fields. [3]
2. Explain what is meant by the term 'unconventional sources' of oil and gas. [4]

ONLINE

Exam tip

The command 'justify your choice' means explaining why you have chosen one option, and why you have rejected other options. Good answers will consider the advantages and disadvantages of both the chosen and rejected options.

Now test yourself and exam practice answers at **www.hoddereducation.co.uk/myrevisionnotes**

The challenge of reducing reliance on fossil fuels

REVISED

Energy efficiency and conservation

Fossils fuels are non-renewable and will run out one day. So every effort must be made to eke out what is left. **Energy conservation** is all about minimising the wastage of energy and using it as efficiently as possible. For example, much can be done in the home by building energy-efficient dwellings or upgrading existing ones to include features such as:

- loft and cavity wall insulation
- double- or triple-glazed windows
- high-efficiency, gas-burning, condensing boilers
- generating energy from small wind turbines and solar panels
- using energy sparingly within the home.

There is much that we can do when it comes to transport, such as walking more, using public transport more and using the car less.

The most important benefit of energy conservation is, of course, that it will reduce carbon emissions. This, in turn, should help to reduce the threat of global warming.

Of the three fossil fuels, gas is the cleanest burning in terms of carbon emissions. So one obvious action here is for countries to move away as much as possible from the use of coal and oil.

Another possible action would be to develop some form of new technology that reduces the carbon emissions resulting from the burning of fossil fuels. There are two possibilities here:

- Carbon scrubbing – done by devices that absorb carbon dioxide exhausts from industrial plants; already used to partially 'clean' natural gas before it is transported.
- Carbon capture and storage – catching the carbon dioxide released by the burning of fossil fuels and then burying it deep underground. This is still at the experimental stage.

Now test yourself

1 Suggest ways in which you could save energy in your home.

TESTED

Now test yourself

2 How will energy conservation reduce carbon emissions?

TESTED

Alternatives to fossil fuels

The simplest way of reducing carbon emissions is to rely less on fossil fuels, and make much greater use of renewable and recyclable energy sources. Most are best used to generate electricity. However, not all countries have the climates and topography needed to create renewable energy resources. Not everyone is happy to see the landscape disfigured by large wind and solar farms. Harnessing water power often means building dams and drowning valleys to create reservoirs.

Of the recyclables, many developed countries need nuclear energy to generate some of their electricity. But there are risks and challenges associated with nuclear power stations.

Now test yourself

TESTED

3 What are the 'risks and challenges' associated with nuclear power stations?

Exam tip

In the Paper 3 exam try to refer to examples in your longer 8- and 12-mark answers, especially if they seem similar to the situation outlined in the Paper 3 Resource Booklet.

The downside of biofuels is that they have to be grown. In many instances, this means on farmland that is needed to produce food for an increasingly hungry world.

Inevitably, the time has come for scientists to invent new energy sources which are both available over much of the globe and environmentally friendly. Perhaps one of the most promising new technologies is the hydrogen fuel cell. This combines hydrogen with oxygen to produce electricity, heat and water for buildings and electric vehicles.

When it comes to energy, all countries today need to:

- reduce their carbon footprint – that is the amount of carbon dioxide released into the atmosphere by the burning of fossil fuels
- broaden their energy mix – that is the combination of different energy sources needed to meet a country's energy demand
- improve their energy security – for a government this is the most important of all and involves satisfying a number conditions (see Figure 9.8).

Revision activity

Make a table listing the costs and benefits of using different sources of renewable and recyclable energy.

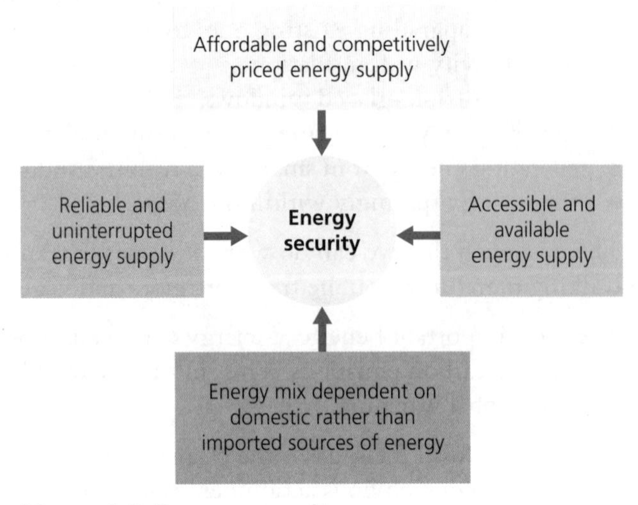

Figure 9.8 Energy security.

Now test yourself

TESTED

1. Why is the energy mix so important?
2. Why is energy security increased by a low dependence on imported energy?
3. What is meant by the term 'carbon footprint'?

Exam practice

1. Explain the downside of using biofuels. [4]
2. Study Figure 9.9, which shows the UK's energy mix. Explain why the UK might lack energy security. [4]

Coal
Natural gas
Nuclear
Renewables
Other

38%
28%
21%
11%
2%

Figure 9.9 The UK's energy mix.

ONLINE

Exam tip

Key words in any definition of energy security are 'affordable', 'reliable', 'accessible' and 'mix'.

Changing attitudes to energy and the environment

REVISED

The link between carbon dioxide emissions and global warming is changing attitudes about energy. More and more people are becoming aware of the need for **sustainable energy**.

Sustainable energy Sources of energy from renewable and recyclable sources which are not diminished by present use and which will be available for the benefit of future generations

Players Individuals, groups or organisations with a stake in a particular issue and the ability to influence outcomes

OPEC The Organization of the Petroleum Exporting Countries: an organisation representing most of the leading oil-producing states

GECF The Gas Exporting Countries Forum: an organisation representing gas-producing states

Major players

The energy scene is dominated by a set of **players** with conflicting vested interests (see Figure 9.10):

- Transnational corporations (TNCs) – most obviously the oil companies, whose aim is to make the energy business a profitable one. This can involve taking risks with the environment.
- International organisations – such as **OPEC** and **GECF** that look after the interests of energy producers (for example, by fixing prices and level of supply).
- Intergovernmental organisations – include those that are particularly concerned with the trade in energy, such as the World Bank and the World Trade Organization.
- National governments – their main aim is energy security and maintaining a sound energy mix.
- Climate scientists – have proven that recent changes in the global climate are due to human activity, particularly the burning of fossil fuels.
- Environmental pressure groups – such as WWF and Greenpeace; their primary concern is the threat of global warming. Their mission is to persuade governments and consumers to shift from fossil fuels to renewables and recyclables.
- Consumers – their priority is a cheap and reliable supply of energy, but a growing number are aware of the environmental costs of cheap energy.

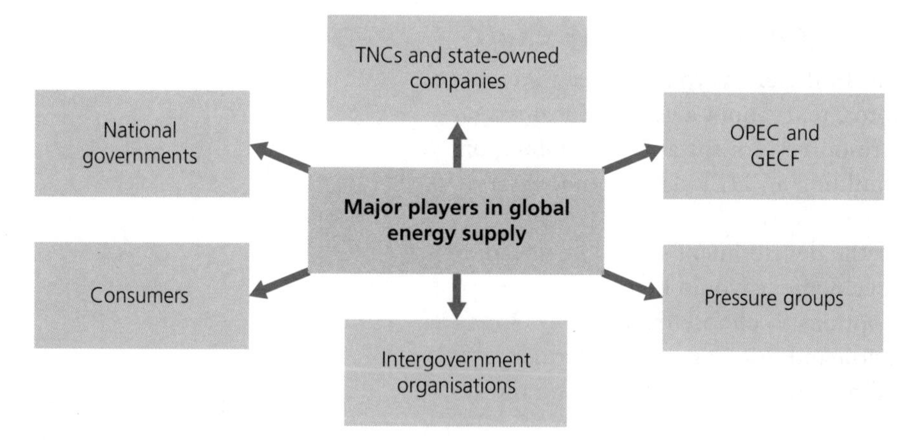

Figure 9.10 Major players in the global supply of energy.

Now test yourself

TESTED

Why do players have different views on energy supply?

Exam tip

Try to remember the six main types of player shown in Figure 9.10.

Changing attitudes

Attitudes to energy and the environmental impacts of energy use are changing because of:

- rising affluence – although this increases energy demand, money is available to invest in cleaner and more sustainable energy
- education – this is raising people's awareness of:
  - the fact that they themselves contribute to the rise in energy demand
  - the environmental concerns of scientists, pressure groups and some world leaders about the impacts of continuing to use fossil fuels
  - the actions that need to be taken to reduce carbon footprints and to achieve a more sustainable supply and use of energy.

But if you live in a developing country, sustainable energy is not quite so attractive. It is more expensive and generally requires access to quite advanced technology.

Now test yourself

How are people being 'educated' about the need for energy conservation?

TESTED

Exam practice

1 Choose **two** of the players in Figure 9.10 on page 143. Compare their attitudes to the energy situation. [3]

2 Study Table 9.3, which sets out three possible options for managing energy in a sustainable way.

Table 9.3 Three energy futures

Option	Strategy
1	Use modern technology to improve the efficiency of burning fossil fuels
2	Develop all the renewable energy sources as quickly as possible
3	Change consumer energy use in homes and transport

Assess the relative merits of these three strategies. [8]

ONLINE

The Paper 3 Making Geographical Decisions exam

Paper 3 is a Decision Making Exam. In the exam you will get a Resource Booklet of data (tables, graphs, photos, text) about a geographical issue. It will combine forests and energy resources, for instance developing oil resources in the Russian taiga, or building an HEP dam in Indonesia's tropical rainforest:

- You will have to decide whether the development should go ahead, be stopped or if some alternative development would be better.
- In the exam there will be three options to choose from.
- You will need to choose one option and justify your decision.
- The last question on Paper 3 will be a 12-mark question using the command 'justify your choice'.
- It is important to understand that none of the options you choose from is 'the right one' – you get marks for how well you justify your choice.

Fz Marc

Susanna Partsch

Franz Marc
1880–1916

TASCHEN
KÖLN LISBOA LONDON NEW YORK OSAKA PARIS

UMSCHLAGVORDERSEITE:
Die großen blauen Pferde (Detail), 1911
Öl auf Leinwand, 104,22 x 180,4 cm
Collection Walker Art Center, Minneapolis,
Gift of the T. B. Walker Foundation,
Gilbert M. Walker Fund, 1942

ABBILDUNG SEITE 1:
Liegender Stier, 1911
Kreide, 11,4 x 15,8 cm
München, Staatliche Graphische Sammlung

ABBILDUNG SEITE 2:
Rotes und blaues Pferd, 1912
Tempera, 26,3 x 43,3 cm
München, Städtische Galerie im Lenbachhaus

UMSCHLAGRÜCKSEITE:
Franz Marc, um 1912

Dieses Buch wurde gedruckt auf 100% chlorfrei gebleichtem Papier gemäß TCF-Norm.

Hohenzollernring 53, D–50672 Köln
Lektorat und Produktion: Brigitte Hilmer, Köln
Umschlaggestaltung: Angelika Muthesius, Köln
Korrekturen: Thomas Paffen, Düsseldorf
Claudia Koppert, Ottersberg
Farbreproduktionen: Repro Color, Bocholt
Schwarzweißreproduktionen: Ceynowa, Köln
Montagen: artcolor, Hamm

Printed in Germany
ISBN 3-8228-0441-X

Inhalt

Auf der Suche nach dem eigenen Weg

Die Beliebtheit von Franz Marc beruht bis heute vor allem auf seinen Tierdarstellungen. Der verschollene *Turm der blauen Pferde* nimmt es an Popularität mit den Blumenbildern Emil Noldes auf. Doch im Gegensatz zu letzterem, der bis ins hohe Alter malte, beschränkte sich die Zeit, in der Marc seine bekannten Gemälde schuf, auf knapp vier Jahre. Zehn Jahre hat es gedauert, bis Franz Marc von der traditionellen Malerei der Münchener Akademie über verschiedene Stilrichtungen zu der Formensprache gelangte, die sich dann als sein Personalstil herauskristallisiert hat. In dieser Zeit - um 1910 - lernte er auch gleichgesinnte Künstler wie August Macke (1887-1914) und Wassily Kandinsky (1866-1944) kennen. Zwischen 1911 und 1914 entstanden nicht nur die wichtigsten Werke; es war auch die Zeit, in der er gemeinsam mit Kandinsky den sogenannten Almanach »Der blaue Reiter« herausgab, die gleichnamigen Ausstellungen organisierte und zu einer bedeutenden Kraft im deutschen Kulturbetrieb wurde. Diese kunstpolitischen Aktivitäten und seine - teilweise kämpferischen - Schriften über die Malerei besitzen einen hohen Stellenwert für eine umfassende Würdigung des Malers. 1914, als Marc gerade begann, sich vom Gegenstand zu lösen, seine ersten abstrakten Gemälde entstanden waren, meldete er sich freiwillig zum Kriegsdienst. Zwei Jahre später fiel er - erst sechsunddreißig Jahre alt - vor Verdun.

Elefant, 1907
Kreide, 41,5 × 32,8 cm
Hamburg, Hamburger Kunsthalle, Kupferstichkabinett

Im Gegensatz zu vielen seiner Künstlerkollegen hat Marc schon früh den Weg eines Künstlers einschlagen können, ohne daß ihm von seiner Familie Steine in den Weg gelegt worden wären. Den üblichen Kampf hatte bereits der Vater ausgefochten, der sich erst nach einem erfolgreich abgeschlossenen Jurastudium den Wunsch erfüllen konnte, sich als Maler ausbilden zu lassen, und später Professor an der Münchener Akademie wurde. Der katholische Wilhelm Marc hatte 1877 die streng calvinistisch erzogene Sophie Maurice geheiratet, die aus einer französischen Familie stammte, ihre Kindheit jedoch hauptsächlich in der Schweiz verbracht hatte. Ihre beiden Söhne Paul und Franz, die 1877 und 1880 in München geboren wurden, wuchsen zweisprachig auf und genossen im katholischen Bayern eine protestantische Erziehung. Auch Wilhelm Marc hatte der katholischen Kirche den Rücken gekehrt und war zum Protestantismus übergetreten.

Franz Marc wuchs in einer strenggläubigen Atmosphäre auf, und der Konfirmandenunterricht hatte auf ihn eine nachhaltige Wirkung. Er blieb mit dem Pfarrer, Otto Schlier, obwohl dieser bereits kurz vor der Konfirmation München verließ, über Jahre in Verbindung und betrachtete ihn als seinen Mentor.

Indersdorf, 1904
Öl auf Leinwand, 40 × 31,5 cm
München, Städtische Galerie im Lenbachhaus

Neben seinem Interesse an der Theologie hatte der junge Marc auch künstlerische und literarische Ambitionen. Er war sich unschlüssig darüber, welcher der drei Neigungen er beruflich den Vorzug geben sollte. Es ist bestimmt auf den Einfluß Otto Schliers zurückzuführen, daß er als Siebzehnjähriger den Wunsch äußerte, Pfarrer zu werden. Ein Jahr später, 1898, kam er jedoch zu dem Entschluß, Philologie zu studieren. 1899 schrieb er sich an der Philosophischen Fakultät in München ein, leistete jedoch vor Beginn des Studiums noch einen einjährigen Militärdienst ab.

Während der Zeit im Lager lernte Marc reiten. Hier mag seine Liebe zu den Pferden gewachsen sein, die ihn sein Leben lang begleitete und die auch in vielen Bildern ihren Niederschlag fand. Als freiwilliger »Einjähriger« stieß Marc seine berufliche Entscheidung erneut um. Nun stand es für ihn fest, daß er Maler werden wollte. In einem Brief erklärte er Otto Schlier, warum diese raschen Wechsel, die ihm wohl erbärmlich und zweifelhaft erscheinen könnten, der einzig gangbare Weg gewesen seien, um einen richtigen Entschluß treffen zu können: »Obwohl ich nämlich mein Leben lang immer schon Künstler war, bin ich doch auch infolge Erziehung und Umgebung und eigener Veranlagung halbwegs Geistlicher und halbwegs Philologe gewesen. Ich hätte wohl als Künstler nie die rechte Ruhe und Sicherheit, wäre ich nicht jenen beiden Idealen zu ihrer Zeit nachgegangen. Jetzt aber weiß ich gewiß, daß ich das Richtige für meine Natur gefunden habe…«[1]

Von 1900 bis 1903 studierte Marc an der Münchener Akademie. Wenn er in der Klasse von Franz von Stuck (1863–1928) gewesen wäre, hätte er dort bereits die späteren Freunde Wassily Kandinsky und Paul Klee (1879–1940) kennenlernen können, doch er wurde Schüler von Gabriel Hackl (1843–1919) und Wilhelm von Dietz (1839–1907), die beide der Münchner Malerschule des 19. Jahrhunderts verpflichtet waren. Er bekam dadurch zwar eine solide Grundausbildung, kam jedoch nicht mit den modernen Strömungen der Zeit in Kontakt. Die wenigen Gemälde, die aus diesen ersten Jahren erhalten sind, zeigen, daß Marc ganz in der Tradition der Münchener Akademie stand. Seine *Moorhütten im Dachauer Moos* (Abb. S. 9) von 1902 sind dafür beredtes Beispiel. Mit feinen Pinselstrichen sind Naturdetails wie die einzelnen Blätter an den Bäumen akribisch ausgeführt. Die Farben sind verhalten. Brauntöne beherrschen das Bild, obwohl das Grün des Mooses einen beträchtlichen Teil des Bildraumes einnimmt.

1903 lud ihn ein begüterter Studienkollege zu einer mehrmonatigen Reise nach Frankreich ein. Vor allem in Paris wurden Marc die Augen für die Gotik, Courbet und die Impressionisten geöffnet.

Als er nach vier Monaten wieder nach München kam, entschloß er sich, nicht an die Akademie zurückzukehren, sondern von nun an selbständig zu arbeiten. Er suchte sich ein eigenes Atelier, doch die Arbeit ging nur mühsam voran.

Nach der Frankreichreise gab Marc die akademische Malweise auf. Seine Farbpalette bestand nun aus helleren, kräftigeren Farben. Wie im Umgang mit der Farbe war auch in der Wahl und Behandlung der Sujets der Einfluß, den die Impressionisten auf ihn ausgeübt hatten, deutlich spürbar. Das zeigt sich z.B. in dem Gemälde *Indersdorf* von 1904 (Abb. S. 6). Mit breitem, flüchtigem Pinselstrich hat er die Wirtschaftsgebäude des Klosters dargestellt. Die Wäschestücke, die auf einem Zaun hängen,

Moorhütten im Dachauer Moos, 1902
Öl auf Leinwand, 43,5 × 73,6 cm
Kochel am See, Franz Marc-Museum
Leihgabe Otto und Etta Stangl

erscheinen als bunte Farbflecken. Im Wasser - einem See oder Teich - spiegeln sich eines der Häuser, das Geländer und das rote Wäschestück. Dabei hat Marc versucht, das Irisieren der Wasseroberfläche im Bild einzufangen.

Mit dieser Annäherung an die Impressionisten begann für Marc eine lange Zeit der Suche nach malerischen Ausdrucksmitteln, mit denen er sich hätte identifizieren können. Diese Jahre waren von einer tiefen Melancholie geprägt. Das mag zum einen an seiner damaligen seelischen Verfassung gelegen haben. Er befand sich aber auch in keiner sehr glücklichen Situation.

Kurz nachdem Marc sein eigenes Atelier bezogen hatte, ging er eine enge Beziehung zu der Malerin und Kopistin Annette von Eckardt ein, die verheiratet war und zwei Kinder hatte. Diese Umstände belasteten beide sehr, wodurch auch die Arbeitsschwierigkeiten Marcs zum Teil erklärt werden können. Nach knapp zwei Jahren lösten sie ihre Verbindung, womit allerdings die beiden nicht weniger unglücklich waren. Marc war mit seiner Arbeit unzufrieden und fühlte sich einsam.

Diesem Gefühl zum Trotz lernte er 1905 den Tiermaler Jean Bloé Niestlé (1884-1942) kennen, mit dem ihn bald eine enge Freundschaft verband und der später ebenso wie Marc in Sindelsdorf lebte. Dieser Kontakt hatte weitreichende Folgen für Marcs späteres Werk.

Niestlé regte Marc dazu an, sein gutes Verhältnis zu Tieren künstlerisch umzusetzen. Dem Tiermaler ging es nicht um eine zoologische Darstellung von Tieren, sondern um den Versuch, sich in das Tier einzufühlen, das Wesen des Tieres in einem Bild einzufangen. Diese Sehweise - die die immer vorhandene Vermenschlichung des Tieres nicht bedenkt

– faszinierte Marc, was auch aus einem Brief hervorgeht, den er kurz nach der ersten Begegnung mit Niestlé schrieb: »Jetzt hat er neben ungezählten Tierstudien einen großen Entwurf: einen zweimeterlangen Rahmen..., auf dem er hundert Stare ... malt. Man glaubt, das Zwitschern und Flügelrauschen zu hören. Und keiner gleicht dem anderen! Jedes Tier hat seinen eigenen Ausdruck.«[2]

Marc hatte zwar schon früher einige wenige Tierbilder gemalt, doch die Begegnung mit Niestlé bedeutete für ihn den Anstoß, die Tiermalerei als Mittel künstlerischen Ausdrucks zu begreifen und zu entwickeln. Ein erstes Beispiel für diese neue Sehweise ist das kleine Bild *Der tote Spatz* von 1905 (Abb. S. 11), das Marc malte, als seine Beziehung zu Annette von Eckardt ein Ende fand.

Auf einem hellen Untergrund und vor einem dunklen Hintergrund, deren beider Brauntöne in der Färbung des Tieres aufgenommen werden, liegt ein kleiner Spatz auf dem Rücken, leblos und starr. Die Bewegung der Pinselstriche, mit denen Unter- und Hintergrund gemalt sind, vergegenwärtigen die Diskrepanz, daß das Tier, welches sich eigentlich in die Lüfte heben und fortfliegen kann, nun an die Erde gebunden ist. Marc hat in diesem Bild offensichtlich auch viele seiner eigenen Gefühle zum Ausdruck gebracht.

Im Herbst desselben Jahres lernte Marc auch die beiden Malerinnen Marie Schnür (geboren 1869) und Maria Franck (1878–1955) kennen. Die elf Jahre ältere Marie Schnür war Lehrerin am Münchner Künstlerinnenverein. Maria Franck lernte an derselben Institution. Mit beiden verband ihn schnell eine enge Freundschaft. Bereits im Februar 1906 fuhr er mit Maria Franck nach Kochel am See, um zu malen. Aus dieser Zeit stammen zwei Porträts seiner späteren Lebensgefährtin und Ehefrau. Im Frühjahr 1906 begleitete Marc seinen Bruder auf den Berg Athos und

Kopf des Vaters auf dem Totenbett, 1907
Kreide, 23,5 × 29,5 cm
München, Staatliche Graphische Sammlung

Der tote Spatz, 1905
Öl auf Holz, 13 × 16,5 cm
Norden, Dr. Erhard Kracht

ging nach seiner Rückkehr erneut nach Kochel. Er begann, die Pferde auf den Weiden zu studieren und versuchte, ihre Bewegungen auf der Leinwand einzufangen. Das wichtigste Beispiel aus dieser Zeit der Freilichtmalerei ist das Gemälde *Zwei Frauen am Berg*, das Marie Schnür und Maria Franck auf einer Wiese zeigt. Von dem Motiv existiert nur noch eine kleine Skizze (Abb. S. 13), da Marc das großformatige Gemälde später zerschnitt. Die hellen, sparsam eingesetzten Farben zeigen, daß Marc sich damals bereits mit der Farbenlehre auseinandersetzte, und der schnelle, breite Pinselstrich zielt nicht auf naturgetreue Wiedergabe ab, sondern auf das Einfangen einer Situation. Darüber hinaus ist dieses Bild auch für seine damaligen Lebensumstände ein wichtiges Zeugnis. Mit Maria Franck verband ihn zu jener Zeit bereits eine sehr enge, intensive Beziehung. Auf dem Gemälde ist sie jedoch in den Hintergrund gerückt und verbirgt ihr Gesicht in der Hand. Marie Schnür, im Vordergrund sitzend, blickt zurück auf die liegende Gestalt. Hier ist nicht nur die Bekanntschaft, die Verbundenheit der beiden Frauen spürbar, die sich bereits länger kannten, sondern auch ihre Konkurrenz.

Marc heiratete nämlich erst einmal Marie Schnür! Die Malerin hatte ein uneheliches Kind, das sie nach der damaligen Rechtslage nur zu sich

Kleine Pferdestudie II, 1905
Öl auf Pappe, 27 × 31 cm
Kochel am See, Franz Marc-Museum,
Eigentum der Bayerischen
Staatsgemäldesammlungen

nehmen konnte, wenn sie verheiratet war. Franz Marc erklärte sich bereit, ihr zu helfen. Ob er sich auch zu ihr stark hingezogen fühlte oder seine Beziehung zu Maria Franck damals schon die intensivere war, läßt sich heute nicht mehr beurteilen. Später hieß es jedoch in einer Eingabe bei Gericht: »Frau Schnür nahm damals das Eheversprechen nur mit dem ausdrücklichen Bemerken an, daß Herr Marc die Freiheit behielte, nach Eingehung der Ehe sich wieder von ihr zu trennen. Dies Bemerken kann zeugenschaftlich bestätigt werden. Frau Schnür wußte damals und in der Folgezeit genau Bescheid über die intimen Beziehungen zwischen Herrn Marc und Frl. Franck... Frau Schnür erklärte, nur um des Kindes willen die Ehe eingehen zu wollen...«[3] Die beiden ließen sich im März 1907 trauen. Marc fuhr noch am Abend nach der Trauung allein nach Paris.

Die Impressionisten, die ihn 1903 nachhaltig beeinflußt hatten, sah er jetzt mit kritischeren Augen. Er bemängelte an ihnen, daß sie zu maßvoll seien, daß ihnen das »Fortissimo« fehle. Marc vermißte bei diesen Malern, die ihrer Kunst eine naturwissenschaftlich-rationale Denkweise zugrunde gelegt hatten, die Emotionalität. Er kritisierte ihre »Sachen« als »bildhaft«, insofern sie nie eine »raum- und seelensprengende« Wirkung hätten, was er für sich als Ziel in Anspruch nahm. Als er dann eine »Riesenkollektion« der Werke von van Gogh und Gauguin sah, war er begeistert. Euphorisch schrieb er an Maria Franck, daß sich seine »schwankende, geängstigte Seele« beim Anblick »dieser wunderbaren Werke endlich beruhigt« hätte.

Zwei Frauen am Berg, Skizze, 1906
Öl auf Leinwand, auf Pappe aufgezogen, 15,5 × 24,7 cm
Kochel am See, Franz Marc-Museum, Eigentum der Bayerischen Staatsgemäldesammlungen

Marc setzte sich mit diesen beiden Künstlern, die als die Überwinder des Impressionismus und Entdecker einer neuen Formensprache galten, intensiv auseinander. Es scheint, daß er – wie etliche andere Maler auch – erst die Stilentwicklungen der letzten Jahrzehnte tätig nachvollziehen mußte, um zu der Malweise zu gelangen, die ihm selber entsprach. So setzte er nun, als er nach München zurückgekehrt war, seine neuen Beobachtungen, die er in Paris gemacht hatte, in seiner Malerei um.

Diese Entwicklung zeigt sich z. B. in *Frau im Wind am Meer* (Abb. S. 14). Das Gemälde entstand im Herbst 1907, als Marc mit Marie Schnür nach Swinemünde zu ihrer Familie gefahren war. Die Farben der Palette wurden noch einmal deutlich heller und kräftiger. Es war ihm vor allem daran gelegen, Bewegung darzustellen. Marc gelang in diesem Gemälde, die Wellen des Meeres und die Gestalt, die gegen den Wind anzulaufen scheint, mit einfachen Mitteln, ohne naturalistische Ausgestaltung und mit wenigen stark kontrastierenden Farben (Blau und Grün) auf der Leinwand festzuhalten.

Von dieser Reise, die aufgrund einer zeitlich offensichtlich sehr begrenzten Annäherung zwischen Marc und seiner Frau stattgefunden hatte, schrieb er an Maria Franck: »Wird mir das Schicksal wohl jemals die Dummheit vergeben, die ich mit dieser Heirat angerichtet habe? Heut komm ich zu Dir und sag: hilf mir! Im übrigen muß ich doch gleich dahinter setzen, daß ich fest überzeugt bin, daß mir dieses eklige Jahr von größtem seelischem Nutzen gewesen ist.«[4]

ABBILDUNG SEITE 14:
Frau im Wind am Meer, 1907
Öl auf Pappe, 25 × 16 cm
Kochel am See, Franz Marc-Museum

ABBILDUNG SEITE 15:
Lärchenbäumchen, 1908
Öl auf Leinwand, 100 × 71 cm
Köln, Museum Ludwig

1908 wurde die Ehe geschieden. Entgegen den Vereinbarungen, die offensichtlich vorher zwischen Franz Marc und Marie Schnür getroffen worden waren, klagte diese auf Ehebruch. Damit verhinderte sie nach dem geltenden Recht eine Heirat zwischen Franz Marc und Maria Franck. Jahrelang dauerte der Kampf, bis es den beiden, die seit 1908 de facto zusammenlebten und sich 1911 in London nach englischem Recht trauen ließen, im Jahre 1913 endlich möglich wurde, nach deutschem Recht zu heiraten.

In diesem Sommer arbeiteten beide in Lenggries. Es entstanden vor allem Baumstudien und Pferdebilder. Maria Franck und Franz Marc arbeiteten zusammen draußen im Freien auf einer Lichtung oder auf den Pferdewiesen. Das *Lärchenbäumchen* (Abb. S. 15) von 1908 zeigt seine Hinwendung zu van Gogh, die in dem Bild des *Eichenbäumchens* ein Jahr später noch stärker zu Tage tritt. Hier in Lenggries begann er, Pferdegruppen auf großformatiger Leinwand zu malen. Dieses Thema beschäftigte ihn noch lange und erfuhr immer neue Variationen.

Diese Bilder von 1908 aus Lenggries und von 1909 aus Sindelsdorf, wie z.B. *Rehe in der Dämmerung* (Abb. S. 17), zeigen, daß Marc immer noch um eine eigene Formensprache rang. Immer stärker reduzierte er die Darstellung auf das Wesentliche, doch konnte er sich noch nicht von der naturalistischen Farbgebung lösen. Erste Anzeichen für seinen Schritt hin zu einer subjektiven Farbenwahl finden sich erst in dem Bild *Akt mit Katze* (Abb. S. 22) von 1910, die er aber in dem Gemälde *Weidende Pferde I* (Abb. S. 25) wieder weitgehend zurücknahm.

Kleines Pferdebild, 1909
Öl auf Leinwand, 16 × 25 cm
Kochel am See, Franz Marc-Museum,
Leihgabe aus Münchner Privatbesitz

Rehe in der Dämmerung, 1909
Öl auf Leinwand, 70,5 × 100,5 cm
München, Städtische Galerie im Lenbachhaus

1909 verbrachte er den Sommer mit Maria Franck erstmals in Sindelsdorf. Ein Jahr später gaben die beiden das Münchner Atelier auf und zogen nach Sindelsdorf um. Geld verdiente Marc zu diesem Zeitpunkt mit seinen Bildern immer noch nicht. Finanziell hielt er sich mit Malkursen und der Vermittlung von Antiquitäten über Wasser, was ihm Kraft und Zeit für seine Malerei raubte. Doch konnte er wenigstens durch die Unterstützung eines Malerkollegen aus der Akademiezeit erreichen, daß die Kunsthandlungen Brakl und Thannhauser in München einige Arbeiten von ihm kauften. Bei Brakl sah ein junger Maler zwei seiner Lithographien und besuchte ihn daraufhin in seinem Atelier. Diese Begegnung zwischen Franz Marc und dem sieben Jahre jüngeren August Macke bedeutete für Marc einen folgenreichen Einschnitt.

Künstlerfreundschaften und Auseinandersetzungen über die Farbe

Im Jahr 1910 vollzog sich eine Wende im Leben von Franz Marc. Er kehrte München endgültig den Rücken und zog sich in die Einsamkeit Oberbayerns zurück. Gleichzeitig gewann er endlich Künstlerfreunde, die auf sein Schaffen einen nicht zu unterschätzenden Einfluß hatten, und engagierte sich in großem Maße für die »neue Kunst«. In seinen Bildern kann diese Entwicklung deutlich nachvollzogen werden.

Der Eindruck, den die Bilder van Goghs und Gauguins auf seiner zweiten Parisreise auf ihn gemacht hatten, wurde durch eine Ausstellung von Werken van Goghs im Dezember 1909 in München noch vertieft. Marc half damals den beiden Kunsthändlern Brakl und Thannhauser beim Hängen der Bilder. Der direkte Umgang mit den Gemälden, der es ihm ermöglichte, die Technik und die Formensprache van Goghs ganz genau zu studieren, führte zu einer konkreten Auseinandersetzung mit dessen Malweise, die in dem Bild *Katzen auf rotem Tuch* (Abb. S. 21) dokumentiert ist. Durch Marcs Briefwechsel mit Maria Franck weiß man, daß er an diesem Gemälde von Ende Dezember bis Anfang Januar arbeitete. Nicht nur die intensive Farbigkeit, sondern vor allem die breiten, kräftigen und bewegten Pinselstriche lassen unwillkürlich an van Gogh denken. Dessen Stilmittel adaptierte Marc vor allem bei der Blumenwiese, die die Katzen im Hintergrund einrahmt. Das Bild fiel bei der Ausstellung, die die Kunsthandlung Brakl im Februar für Marc ausrichtete, völlig aus dem Rahmen. Es widersprach auch den Erwartungen, die man haben mochte, wenn man das Plakat sah, das Marc für die Ausstellung entworfen hatte (Abb. S. 19). Auch dort sind zwei Katzen dargestellt, deren Bewegungen Marc meisterlich eingefangen hatte. Doch fehlt ihnen noch die Beschränkung auf das Wesentliche, auf die Marc nun in verstärktem Maße hinarbeitete.

Bei einem anderen Gemälde aus dem Jahr 1910, dem *Akt mit Katze* (Abb. S. 22), verarbeitete Marc wiederum andere Eindrücke. Hier haben Henri Matisse (1869–1954) und Paul Cézanne (1839–1906) Pate gestanden. Die Behandlung der Figur erinnert an Cézanne, während die Farbigkeit eher auf den Einfluß von Matisse zurückzuführen ist. Marcs damalige Auseinandersetzung mit den Spektralfarben und seine Versuche mit einer subjektiven Farbenwahl finden in diesem Gemälde ihren Niederschlag. Allerdings mied er große Farbflächen, so daß das Bild unruhig wirkt und die einzelnen Teile auseinanderzufallen scheinen.

Betrachtet man die Gemälde der beiden folgenden Jahre, so wird die Suche nach dem eigenen Stil deutlich. Marc war aber offensichtlich noch nicht ganz imstande, die Ideen, die ihm vorschwebten, in Bilder

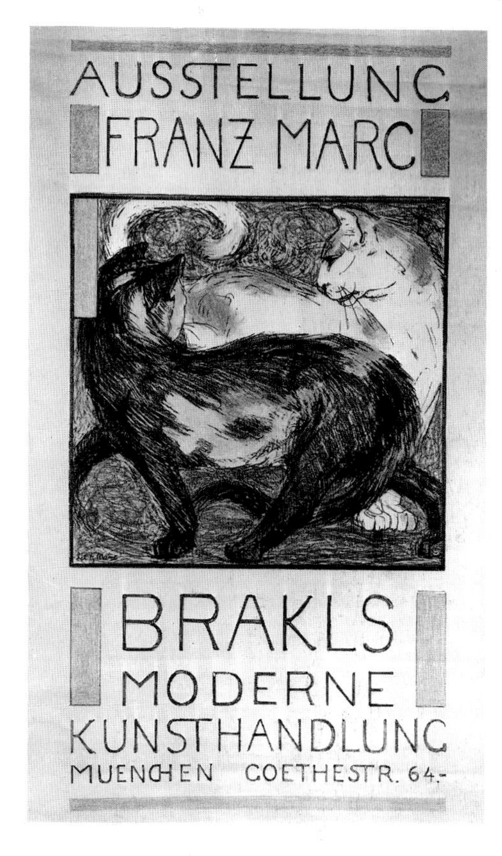

Zwei Katzen, 1909/10
Plakat für die Ausstellung Franz Marc in Brakls Moderner Kunsthandlung München, Februar 1910
Lithographie, 40,5 × 42 cm
München, Städtische Galerie im Lenbachhaus

Blaues Pferd I, 1911
Öl auf Leinwand, 112,5 × 84,5 cm
München, Städtische Galerie im Lenbachhaus

umzusetzen. Die Orientierung an Stilmitteln derjenigen Maler, die ihn faszinierten, darf nicht als ein Kopieren mißdeutet werden. Es waren tastende Versuche, die eigene Vorstellungswelt darzustellen. Im *Akt mit Katze* ist ihm dies offensichtlich nicht ganz gelungen.

So scheint es verständlich, daß er in *Weidende Pferde I* (Abb. S. 25) die subjektive Farbenwahl wieder stärker zurücknahm. In ihren Bewegungen, in ihrem »Ausdruck« reduzierte er die Darstellung der Pferde zwar auf das Wesentliche, die naturalistische Abbildungsweise ist also zugunsten der Betonung eines Wesenszuges, einer Handlung aufgegeben. Mit den hellen, kontrastarmen Braun- und Gelbtönen, denen wenig Blau und viel Weiß beigegeben ist, griff Marc jedoch auf seine frühere, eher naturnahe Verwendung der Farben zurück.

Um so erstaunlicher wirkt in dieser Folge das *Pferd in Landschaft* (Abb. S. 30) aus dem selben Jahr. Große Farbflächen dominieren das Bild. Das rotbraune Pferd mit seiner blauschwarzen Mähne im Vordergrund blickt in das Bild hinein auf ein hellgelbes Feld, welches von grünen Farbflecken durchbrochen wird. Nur durch das Pferd werden wir angeleitet, in diesen Farbflächen ein Weizenfeld und grüne Büsche zu erkennen. Mit seinen kräftigen, kontrastreichen Farbfeldern strahlt das Gemälde eine Ruhe und Harmonie aus, wie sie vorher im Œuvre Franz Marcs nicht zu beobachten sind.

Nach fünf Jahren des Suchens, der Auseinandersetzung mit den Impressionisten, mit den Vätern des Expressionismus, wie van Gogh und Gauguin immer wieder genannt werden, und schließlich mit den »Fauves«, den Wilden um Matisse, hatte Marc nun innerhalb eines Jahres in seiner Malerei eine stilistische Entwicklung vollzogen, die zur Ausprägung seines eigentümlichen Stils geführt hatte. Das wird deutlich, wenn man zwei thematisch ähnliche Bilder wie *Rehe in der Dämmerung* (Abb. S. 17) von 1909 und *Rehe im Schnee* (Abb. S. 27) von 1911 miteinander vergleicht. Auf beiden Gemälden sieht man zwei Rehe. Das eine hat jeweils den Kopf nach vorne gebeugt, das andere hat ihn erhoben und blickt witternd hinter sich. Die Tiere befinden sich an einem nicht genau bestimmbaren Ort, einmal auf einer Wiese, das andere Mal im Schnee. Dieser Beschreibung zufolge müßte es sich um zwei sehr ähnliche Bilder handeln. Farbgebung und Malweise lassen jedoch aus dem sehr ähnlichen Sujet zwei in ihrer Struktur völlig unterschiedliche Gemälde werden. In dem früheren Bild herrschen gebrochene Weiß- und Brauntöne vor. Die Tiere und der Hintergrund sind aus kurzen, breiten Pinselstrichen aufgebaut. Die Grasbüschel scheinen aus kleinen, flackernden Flammen zu bestehen, durch die eine gewisse Unruhe aufkommt.

Im Gegensatz dazu ist auf dem späteren Gemälde der Pinselstrich regelmäßig geschichtet und zu glatten Flächen vertrieben. In große, weiße Schneeberge bohren sich blaugrüne Löcher hinein. Sie bilden die Kulisse für die rötlichgelben Rehe, deren Körperbau nur noch angedeutet ist. Die Vereinfachung und die Beschränkung der Formen auf das Wesentliche, die auch in dem früheren Bild schon zu beobachten war, ausgeführt in großen, reinen Farbflächen, läßt diese Darstellung ruhiger und harmonischer erscheinen. Die Bildauffassung ist eine andere geworden. Während beim ersten Bild die vordergründige Darstellung überwiegt, hatte Marc nun einen Weg gefunden, die »innere, geistige Seite der Natur« zu malen.

Katzen auf rotem Tuch, 1909/10
Öl auf Leinwand, 50,5 × 60,5 cm
Privatbesitz

Der beschriebene Prozeß: die freie Verwendung der Farben, die Veränderung der formalen Strukturen eines Bildes und auch das Selbstbewußtsein, mit dem Marc jetzt seine Malerei vorantrieb, sind maßgeblich auf die Kontakte zurückzuführen, die er inzwischen geknüpft hatte. Andererseits setzten diese Kontakte seine künstlerische Entwicklung voraus, Marcs Ausstrahlung und die Anregungen, die er erhielt, bedingten sich gegenseitig.

So ist das erste Zusammentreffen zwischen Macke und Marc auf die Initiative Mackes hin zustande gekommen, der angesichts einiger weniger Graphiken Marcs im Januar 1910 sofort den Wunsch verspürte, den Maler kennenzulernen. Die erste Begegnung zwischen Franz Marc und August Macke ist sowohl von dem Münchener Maler in einem Brief an Maria Franck als auch von Mackes Vetter Helmuth Macke beschrieben worden, so daß die Details bekannt sind.

Der damals dreiundzwanzigjährige Macke, der eigentlich in Bonn lebte, wohnte für ein Jahr mit seiner jungen Frau zusammen in Tegernsee. Anfang Januar des Jahres 1910 fuhr er mit seinem Vetter Helmuth, ebenfalls Maler, und dem Vetter seiner Frau, Bernhard Koehler jun., nach München, wo sich die drei in verschiedenen Kunsthandlungen umsehen wollten. Koehler stammte aus Berlin und war der Sohn des reichen Fabrikanten und Mäzens Bernhard Koehler sen..

August Macke und seine beiden Begleiter entdeckten in der Kunsthandlung Brakl zwei Lithographien von Marc, die sie derart begeisterten, daß sie Brakl um die Adresse dieses Künstlers baten. Kurz darauf standen sie in seinem Atelier in der Schellingstraße. Für Marc war diese Begegnung um so wichtiger, als er das Gefühl hatte, zum erstenmal mit Leuten zusammenzutreffen, die dachten und fühlten wie er. Vor allem

zu August Macke fühlte er sich hingezogen. Bereits wenige Wochen später fuhr er zusammen mit Maria Franck für einige Tage nach Tegernsee, wo die Freundschaft der beiden Maler besiegelt wurde. Wieder zurück in München, schrieb Marc an Macke: »Ich halte es für einen wirklichen Glücksfall, endlich einmal Kollegen von so innerlicher, künstlerischer Gesinnung getroffen zu haben, – rarissime! Wie werde ich mich freuen, wenn es uns einmal gelingen sollte, Bild an Bild nebeneinanderzustellen.«[5] Zu dieser Zeit hatte Marc noch keine Gemälde, sondern lediglich Zeichnungen des Freundes sehen können.

Die Freundschaft der beiden Maler ist durch einen intensiven Briefwechsel dokumentiert, in welchem sich ihre Beziehung spiegelt. Sie sprachen in ihren Briefen über ihre Arbeitsweisen, kritisierten in einer für uns heute erstaunlichen Offenheit die Bilder des anderen, wenn sie ihnen nicht zusagten, und blieben einander trotz aller Meinungsverschiedenheiten in künstlerischen und kunstpolitischen Fragen persönlich eng verbunden.

Koehler jun. hatte bei Brakl nicht nur eine Lithographie von Marc gekauft, sondern den Galeristen auch gebeten, seinem Vater einige Bilder Marcs nach Berlin zur Ansicht zu schicken. Koehler sen. kam daraufhin bereits Ende Januar nach München und besuchte Marc in seinem Atelier. Dort entdeckte er auch das für Marc so wichtige Bild des *Toten Spatzen* (Abb. S. 11), das immer auf dessen Schreibtisch stand. Marc hatte immer wieder erklärt, daß es unverkäuflich sei. Als Koehler dafür jedoch hundert Mark bot, überwog die Einsicht in die finanziellen Schwierigkeiten, in denen er sich befand, und er verkaufte das Bild. Es wurde der Grundstein für die große Sammlung, die Bernhard Koehler später von Marc besaß und die sich heute bis auf die Teile, die im zweiten Weltkrieg verbrannten, in der Städtischen Galerie im Lenbachhaus in München befindet. Kurz nachdem Marc ganz nach Sindelsdorf gezogen war, fuhr er nach Berlin und lernte dort die Sammlung Koehlers kennen. Zum erstenmal sah er nun auch Gemälde von August Macke. Wieder in Sindelsdorf, schrieb er an den Freund: »... und alles, was sonst von Ihnen dort hängt, verrät den famosen Maler, den ich angesichts Ihrer Skizzenbücher vermutete und erhoffte... Die Koehlersammlung enthält ja köstliche Stücke. Aber eine Neugestaltung ist dringend nötig, wenn sie als Ganzes einem Freude machen soll. Wir wollen uns beide recht bemühen, dass die Sache in fünf Jahren anders ausschaut.«[6]

Marc konnte hier so selbstsicher von sich sprechen, weil Koehler mit ihm die Abmachung getroffen hatte, ihm – erst einmal auf ein Jahr begrenzt – monatlich zweihundert Mark zu zahlen und als Gegenwert dafür Bilder nach seiner Wahl zu erhalten. Außerdem hatte Koehler bereits in München aus der Ausstellung bei Brakl einige Bilder erworben, unter anderem die *Katzen auf rotem Tuch* (Abb. S. 21). Damit war Marc frei von finanziellen Sorgen. Das ist wahrscheinlich auch ein Grund für die nun einsetzende schnelle Entwicklung in seiner Malerei, die oben beschrieben wurde. Marc konnte endlich ungestört arbeiten, sich ganz auf seine Bilder konzentrieren. Den Sommer über malte er in Sindelsdorf. Der Kontakt zu Macke, der erst im Spätherbst wieder nach Bonn zurückkehrte, wurde enger. Man besuchte sich, traf sich in München, tauschte Ideen, Gedanken und auch Bücher aus. Im Herbst desselben Jahres bekam Marc endlich Kontakt zu der Neuen Künstlervereinigung München.

Akt mit Katze, 1910
Öl auf Leinwand, 86,5 × 80 cm
München, Städtische Galerie im Lenbachhaus

August Macke
Bildnis Franz Marc, 1910
Öl auf Pappe, 50 × 39 cm
Berlin, Staatliche Museen Preußischer Kulturbesitz, Nationalgalerie

Diese Vereinigung, die im Januar 1909 gegründet worden war und der unter anderem Wassily Kandinsky, Alexej von Jawlensky (1864–1941), Gabriele Münter (1877–1962) und Marianne von Werefkin (1860–1938) angehörten, zeigte im Dezember desselben Jahres zum ersten Mal in der Galerie Thannhauser ihre Werke. Marc war damals zwar beeindruckt, hatte jedoch keine Veranlassung, Kontakte zu den Künstlern und Künstlerinnen zu suchen. Er sah offensichtlich noch nicht die Gemeinsamkeiten, die für ihn einen Austausch hätten sinnvoll erscheinen lassen.

Die zweite Ausstellung der Gruppe fand im September 1910 statt und rief polemische Kritiken des Publikums hervor. Marc, der die Ausstellung in den ersten Tagen besuchte, war begeistert und reagierte auf die Kritik, die ihm offensichtlich während des Ausstellungsbesuches zu Ohren gekommen war, mit einem langen Schreiben. Darin erklärte er die negative Reaktion des Publikums mit dessen konservativen Sehgewohnheiten (»es sucht Staffeleikunst«) und versuchte, das Neue an diesen Bildern in Worte zu fassen: »Die völlig vergeistigte und entmaterialisierte Innerlichkeit der Empfindung, der im ›Bild‹ beizukommen unsre Väter... nie auch nur versuchten. Dies kühne Unterfangen, die ›Materie‹, an der sich der Impressionismus festgebissen hat, zu vergeistigen, ist eine notwendige Reaktion... Was bei diesem neuen, das die ›neue Künstlervereinigung‹ macht, uns so aussichtsreich erscheint, ist, daß ihre Bilder neben ihrem auf's Höchste vergeistigten Sinn höchst wertvolle Exempel für Raumaufteilung, Rhythmus und Farbentheorie enthalten.« Nachdem Marc bei einzelnen Künstlern die Überwindung der traditionellen Malerei nachgewiesen hatte, fuhr er fort: »Ihre konsequent durchgeführte Aufteilung der Fläche, die geheimnisvollen Linien des einen, der Farbenklang des andern sucht geistige Stimmungen auszulösen, die mit der Materie des Dargestellten wenig zu thun haben aber einer neuen, sehr vergeistigten Ästhetik den Boden bereiten...«, und schloß mit den Worten: »Wer Augen hat, muß hier den machtvollen Zug der neuen Kunst sehen.«[7]

Diese Zitate zeigen Marcs Gedanken über die Funktion, die für ihn die Kunst haben sollte. Das »Geistige«, ein Begriff, den er nie genauer definierte, spielt für ihn dabei eine wesentliche Rolle. Interessant sind diese Gedanken vor allem insofern, als Kandinsky damals bereits die Arbeiten zu seinem programmatischen Werk »Über das Geistige in der Kunst« abgeschlossen hatte. Das Buch erschien dann erst durch die Vermittlung Marcs Ende 1911. Die durchaus ähnlichen kunsttheoretischen Ansätze erklären die Begeisterung, mit welcher Marcs Schrift bei der Künstlervereinigung aufgenommen wurde, und auch den sehr bald zustande kommenden intensiven Kontakt.

Marc schickte seine Kritik dem Verleger Reinhard Piper, mit dem er seit einigen Monaten in Verbindung stand, und bat ihn, diese weiterzuleiten. So gelangte sie in die Hände der Künstler. Bald darauf erschien in den »Münchner Neusten Nachrichten« eine vernichtende Kritik. Ihr konnte man entnehmen, »daß die Mehrzahl der Mitglieder und Gäste der Vereinigung unheilbar irrsinnig ist« und daß die Ausstellung »konzentrierter Unsinn« sei. Die Vereinigung bat Marc um Erlaubnis, seine Schrift neben der aus den »Münchner Neusten Nachrichten« in einer Broschüre abzudrucken, um diese auf den weiteren Stationen der Ausstellung zu verkaufen. So war ein Kontakt hergestellt.

Weidende Pferde I, 1910
Öl auf Leinwand, doubliert, 64 × 94 cm
München, Städtische Galerie im Lenbachhaus

Marc lernte bald einige Mitglieder kennen (Adolf Erbslöh und Alexander Kanoldt, etwas später auch Marianne von Werefkin und Alexej Jawlensky). Im Januar 1911 traf er dann schließlich auch mit Kandinsky zusammen. Begeistert schrieb er an Maria Franck: »Kandinsky übertrifft alle ... an persönlichem Reiz; ich war völlig gefangen ... Ach, wie freue ich mich, später mit Dir mit diesen Menschen zu verkehren, Du wirst Dich sofort wohl fühlen ...«[8]

Im Februar 1911 wurde Marc zum Mitglied und gleichzeitig zum 3. Vorsitzenden der Vereinigung gewählt. Das war der Beginn für eine eingehende Beschäftigung mit der damaligen Kunstpolitik, die in den folgenden Jahren sehr viel Raum und Zeit in seinem Schaffen einnehmen sollte.

Marc hatte sich schon seit geraumer Zeit mit der Farbenlehre befaßt, dabei aber die alten Regeln von der Benutzung der Komplementärfarben nicht in Frage gestellt. Im Winter 1910/11 griff er diese Studien wieder auf, jedoch nun mit dem Bemühen um eigene Ideen und Vorstellungen. Den Anstoß dazu hatte er von zwei Seiten erhalten.

Im Dezember 1910 berichtete er Maria Franck von seinen Farbexperimenten und erwähnte die Bemerkung Marianne von Werefkins, die Deutschen machten alle den Fehler, das Licht für die Farbe zu nehmen, obwohl die Farbe etwas ganz anderes sei, was ihn angestachelt habe, die

Farbphänomene genau zu untersuchen. Andererseits hatten ihn auch Mackes Bilder zu dieser Auseinandersetzung angeregt. Marcs und Mackes Briefwechsel in diesem Winter dokumentiert einerseits die Gedanken Marcs, zeigt aber auch, daß beide von den Erkenntnissen des jeweils anderen profitierten. So kann man bei ihnen nicht von einem Spiritus rector reden, sondern von wiederholter gegenseitiger Befruchtung, bis sich ihre Auffassungen von Malerei soweit auseinanderentwickelt hatten, daß nur noch eine gegenseitige Akzeptanz möglich war. Doch das war zu einem späteren Zeitpunkt.

Im Dezember 1910 begann ihr Erfahrungsaustausch über die Farbe. Macke hatte sich gerade mit der Parallelität von Farbe und Musik beschäftigt und ordnete der Farbe Blau traurige Klänge zu, dem Gelb heitere und dem Rot brutale. Er schrieb an den Freund: »Die Grenzen von Gelb, Rot und Blau verschmelzen zu Orange, Violett, Grün, wobei das Hellerwerden dem Höhersteigen der Klaviertöne entspricht, wobei die Masse der Oktaven des Klaviers... der Zahl der konzentrischen Kreise entspricht.«[9] Macke, der sich einen Farbring offenbar mit den Primär- und Sekundärfarben hergestellt hatte, bat Marc in demselben Brief, ihm seine Gedanken dazu mitzuteilen. Er beschäftige sich mit der Farbtheorie, weil er es für wichtig halte, »allen Malgesetzen auf den Grund zu gehen, besonders die modernsten mit den ältesten zu verknüpfen, um naiv mit der Kunst sich ausdrücken zu können«.

Marc antwortete wenige Tage später. Den Farbenkreis lehnte er ab, weil sich die Komplementärfarben immer nur in der Mitte träfen und man sie nie nebeneinander sehen könne. Anschließend gab er dem Freund Auskunft über seine Vorstellung von den Primärfarben und dem jeweiligen Charakter, den sie für ihn verkörperten: »Blau ist das männliche Prinzip, herb und geistig. Gelb das weibliche Prinzip, sanft, heiter und sinnlich. Rot die Materie, brutal und schwer und stets die Farbe, die von den anderen beiden bekämpft und überwunden werden muß!«

Nachdem er auf die Sekundärfarben (Grün, Violett und Orange), die sich aus der Mischung je zweier Primärfarben ergeben, und ihr Verhältnis zu den Komplementärfarben eingegangen war – »Blau und Orange, ein durchaus festlicher Klang« –, kam er noch einmal auf das Verhältnis der Primärfarben untereinander zu sprechen: »Ich werde trotz aller Spektralanalysen den Malerglauben nicht los, dass Gelb (das Weib!) der Erde Rot näher steht, als Blau, das männliche Prinzip.«

Er betonte, daß er die Farben nicht gleich werte und daß er das Verhältnis der Komplementärfarben zueinander als viel weniger kompliziert empfinde als das der Farbmassen: »Daß Blau sich auf Orange stützt, ist nicht schwer sich einzuprägen, aber welche Masse Blau sich in jedem einzelnen Falle neben Orange stellen darf, – da liegt der Has' im Pfeffer. Das lassen die Theorien verflucht aus...« Dem Zusammenhang zwischen Musik und Farben stand er skeptisch gegenüber, bat aber den Freund darum, es ihm bei seinem nächsten Besuch vorzuführen.

Es ist offensichtlich, daß beide Maler gerade in dieser Zeit darum bemüht waren, »ihren« Stil zu finden, denn sie tauschten sich über die diversen Farbtheorien und Kompositionsschemata in langen, ausführlichen Briefen aus. Marc las etliche Bücher über Farbtheorie und bemerkte erstaunt: »Überrascht bin ich vor allem über die unglaublich theoretischen Widersprüche, die über diese paar Spektralfarben unter den Gelehrten herrschen. Aber ich habe interessante Dinge gefunden.«

Rehe im Schnee, 1911
Öl auf Leinwand, 84,7 × 84,5 cm
München, Städtische Galerie im Lenbachhaus

In einem weiteren Brief vom 14. Januar 1911 teilte er Macke wesentliche Gedanken mit, die sich direkt auf seine Malerei auswirkten. Daran läßt sich der Prozeß erkennen, in dessen Verlauf er die Malweise entwickelte, die den wichtigsten Teil seines künstlerischen Schaffens ausmacht. Ein wichtiges Ereignis in diesem Zusammenhang war der Besuch eines Konzertes von Arnold Schönberg (1874–1951) gemeinsam mit den Mitgliedern der Künstlervereinigung, bei dem er mit der Zwölftonmusik konfrontiert wurde. »Kannst Du Dir eine Musik denken, in der die Tonalität (also das Einhalten irgend einer Tonart) völlig aufgehoben ist? Ich mußte stets an Kandinskys grosse Komposition denken, der auch keine Spur von Tonart zuläßt... Schönberg geht von dem Prinzip aus, dass die Begriffe Konsonanz und Dissonanz überhaupt nicht existieren. eine sogenannte Dissonanz ist nur eine weiter auseinanderliegende Konsonanz.« Diese Idee beschäftigte Marc im Zusammenhang mit den Farben. Er folgerte für sich daraus, daß man sich nicht an die alten Farbgesetze halten müsse: »Es ist durchaus nicht erforderlich, dass man die Komplementärfarben wie im Prisma nebeneinander auftauchen lässt, sondern man kann sie so weit man will ›auseinanderlegen‹. Die partiellen Dissonanzen, die dadurch entstehen, werden in der Erscheinung des ganzen Bildes wieder aufgehoben, wirken konsonant (harmonisch), sofern sie in ihrer Ausbreitung und Stärkegehalt komplementär sind.«

Marc sah nun zwar auch ein Verhältnis zwischen Malerei und Musik, er verglich allerdings nicht die Töne mit den Farben, wie Macke dies tat, sondern eher die Farbenlehre mit der Harmonielehre und sah, nachdem er die Veränderungen in der einen erlebt hatte, die Möglichkeit, auch die andere neu zu strukturieren.

Die Verbindung von Malerei und Musik, die beide Maler suchten, erklärt sich auch durch die große Rolle, die die Musik in den beiden Familien spielte. Sowohl Lisbeth Macke als auch Maria Franck beherrschten das Klavierspiel und beschäftigten sich mit den modernen Musikströmungen. So besorgte Maria Franck Noten von der Schönbergschen Musik, um sie im Spiel besser nachvollziehen zu können. Daraus zog dann auch Franz Marc Gewinn. Er erfuhr durch das Schönbergkonzert eine ungeheure Bestätigung in seinem Gefühl, daß es nur möglich sei, zu einer neuen Malerei zu gelangen, wenn man die geistigen Grundlagen, die bis dahin das Weltbild des Mitteleuropäers bestimmt hätten, überwände: »... unsere Ideen und Ideale müssen ein härenes Gewand tragen, wir müssen sie mit Heuschrecken und wildem Honig nähren und nicht mit Historie, um aus der Müdigkeit unsres europäischen Ungeschmacks herauszukommen.«

Das waren große Worte, die einzulösen wohl bis heute nicht gelungen ist. Marc jedenfalls hat sich – ob bewußt oder unbewußt – durchaus alter Formen und Inhalte bei seiner Malerei bedient.

Macke, der selten lange Briefe schrieb, ging auf die programmatischen Briefe Marcs nicht ein, bat jedoch um genauere Erklärungen, wie Marc mit dem Prisma umgehe. Die Antwort Marcs ist insofern interessant, als sie seinen Umgang mit den Farben anschaulich darstellt. Marc gebrauchte das Prisma, um seine »gemalten Farben in ihrem Nebeneinander auf Reinheit der Wirkung zu überprüfen«. Im Prisma konnte er dies besser nachvollziehen als mit dem bloßen Auge. Die praktische Anwendung beschrieb er Macke anhand des Bildes *Liegender Hund im Schnee* (Abb. S. 29), das er offensichtlich gerade gemalt hatte:

Liegender Hund im Schnee, 1910/11
Öl auf Leinwand, 62,5 × 105 cm
Frankfurt, Städelsches Kunstinstitut

»Ich malte meinen Russi liegend auf einem Schneefeld; ich machte den Schnee rein weiss mit rein blauen Tiefen; den Hund schmutziggelb. Im Prisma erschien das Gelb trübgrau, der ganze Hund von den tollsten Farbenringen eingerandet. Ich machte nun etappenweise den Hund ›reinfarbiger‹ (hellgelb); mit jedemmal, mit dem die Farbe reiner wurde, verschwanden die farbigen Ränder am Hund immer mehr, bis endlich ein reines Farbverhältnis zwischen dem Gelb, dem kalten Weiss des Schnees und dem Blau darin hergestellt war. Ferner muß die Masse blau gegenüber dem reinen aber lichtschwachen Gelb des Hundes sich nicht zu stark ausbreiten, um noch komplementär (d.h. berechtigt, »organisiert«) zu bleiben.«

Diese Beschreibung zeigt die Arbeitsweise Marcs sehr anschaulich und erklärt auch, warum das Nebeneinandersetzen reiner Farben gar nicht so einfach ist, wie es scheint, wenn man die Bilder betrachtet. Gerade für die damaligen Maler, die aus einer Schule kamen, die sie gelehrt hatte, mit gedeckten Farben umzugehen, bedeutete die Verwendung von reinen, klaren Farben langwierige Experimente und Lernprozesse.

Direkte Verbindungen zwischen den oben zitierten Briefstellen und den Gemälden erkennt man z.B. in den Bildern *Blaues Pferd I* (Abb. S. 18), *Der Stier* (Abb. S. 32), *Die gelbe Kuh* (Abb. S. 40) und natürlich *Liegender Hund im Schnee* (Abb. S. 29), bei dem sein Russi als Modell gedient hatte.

Im Jahr 1911 verstärkte sich auch der gerade geknüpfte Kontakt zu Kandinsky. Der Maler hielt sich oft mit Gabriele Münter in deren Haus

Pferd in Landschaft, 1910
Öl auf Leinwand, 85 × 112 cm
Essen, Museum Folkwang

in Murnau auf, und so konnte man sich entweder zu Fuß oder mit dem Fahrrad besuchen. Der Briefwechsel zwischen Marc und Kandinsky dokumentiert die gemeinsame Arbeit und auch die Verbundenheit, die zwischen den beiden Malern bestand. Er zeigt in seiner Sachlichkeit aber auch, daß zwischen ihnen nie die Herzlichkeit und Vertrautheit aufgekommen ist, die im Briefwechsel mit Macke auch bei den heftigsten Auseinandersetzungen immer die Oberhand behielt. Die neuen Verbindungen brachten große Veränderungen im Leben von Marc mit sich. Obwohl er in der Abgeschiedenheit von Sindeldorf lebte, wurde er jetzt kunstpolitisch aktiv, was einen gut Teil seiner Zeit in Anspruch nahm. Er erschloß Kandinsky die Möglichkeit, sein Manuskript »Über das Geistige in der Kunst« bei Reinhard Piper drucken zu lassen. Es erschien Ende des Jahres 1911. Und er war maßgeblich an der Reaktion auf den »Vinnen-Protest« beteiligt.

Der Worpsweder Maler Carl Vinnen hatte wegen des Ankaufs eines van-Gogh-Bildes durch die Kunsthalle Bremen im April 1911 einen Aufruf gegen die Überfremdung deutscher Kunst gestartet, dem sich viele

»Am anderen Morgen wanderte ich zu Kandinsky! Die Stunden bei ihm gehören zu meinen denkwürdigsten Erfahrungen. Er zeigte mir viel ... im ersten Moment fühle ich die große Wonne seiner starken, reinen, feurigen Farben, und dann beginnt das Gehirn zu arbeiten; man kommt nicht los von diesen Bildern und wenn man fühlt, daß einem der Kopf zerspringt, wenn man sie ganz auskosten will ...«

Franz Marc, 10. 2. 1911

Hocken im Schnee, 1911
Öl auf Leinwand, 79,5 × 100 cm
Kochel am See, Franz Marc-Museum,
Nachlaß Franz Marc

und auch namhafte Künstler angeschlossen hatten, unter anderem Thomas Theodor Heine, Franz von Stuck, Wilhelm Trübner und Paul Schultze-Naumburg, der dann im Nationalsozialismus mit seiner »Rasse-Kunst-Theorie« durchschlagenden Erfolg hatte. Erstaunlicherweise gehörte auch Käthe Kollwitz zu den Unterzeichnern.

Die Kampfschrift »Ein Protest deutscher Künstler« erschien erst auszugsweise in deutschen Tageszeitungen, dann auch als Broschüre. Marc und Kandinsky hatten unabhängig voneinander den Plan gefaßt, etwas dagegen zu unternehmen, und organisierten daraufhin gemeinsam die Schrift »Im Kampf um die Kunst«, in der berühmte Museumsdirektoren, Kunsthistoriker und Künstler Stellung bezogen. Bereits im Sommer erschien »Im Kampf um die Kunst« bei Piper.

Marc selber schrieb in seinem Beitrag: »Ein starker Wind weht heute die Keime einer neuen Kunst über ganz Europa und wo gutes, unverbrauchtes Erdreich ist, geht die Saat auf nach natürlichem Gesetz. Der Ärger einiger Künstler der deutschen Scholle, dass gerade Westwind geht, wirkt wirklich komisch. Sie bevorzugen Windesstille. Den Ostwind

Der Stier, 1911
Öl auf Leinwand, 101 × 135 cm
New York, Solomon R. Guggenheim Museum

mögen sie nämlich auch nicht, denn von Russland her weht es denselben neuen Samen. Was ist da zu machen? Nichts. Der Wind fährt, wohin er will. Der Same stammt aus dem Reichtum der Natur; und selbst wenn Ihr ein paar Pflänzchen mit Füssen tretet oder ausreisst, so macht das der Natur gar nichts aus. Es ist nur etwas unkollegial und verrät auch eine traurige Anschauung über Kunst. Es gibt nur einen Weg der Verständigung: den ehrlichen Vergleich.«[10]

Es war das erstemal, daß Franz Marc eine solch groß angelegte Aktion in die Hände nahm. Briefe mußten geschrieben, Bittgänge zu einzelnen Leuten gemacht werden. Die Verhandlungen mit Reinhard Piper waren hauptsächlich sein Werk. Im Dezember traten Kandinsky, Marc und Münter aus der Künstlervereinigung aus. Der Anlaß ist kaum der Rede wert, es hätte auch ein anderer sein können: Kandinsky hatte ein Bild für die nächste Ausstellung eingereicht, das zu groß war, um juryfrei in die Ausstellung zu gelangen. Es wurde natürlich ausjuriert.

Kandinsky hatte bereits im Januar des Jahres wegen prinzipieller Meinungsverschiedenheiten unter den Mitgliedern den Platz des ersten Vorsitzenden zur Verfügung gestellt. Im August schrieb Marc an Macke, wie sehr ihn die neuen Arbeiten Kanoldts deprimiert hätten, und vermutete (zusammen mit Kandinsky), daß es bei der nächsten Jury zum

Mädchen mit Katze II, 1912
Öl auf Leinwand, 71,5 × 66,5 cm
Privatbesitz

Bruch kommen werde. Es sei nur unsicher, ob Jawlensky und Werefkin im entscheidenden Moment mitziehen würden, so daß bei einer Kampfabstimmung die anderen die Vereinigung verlassen müßten. Marcs Befürchtungen bewahrheiteten sich. Obwohl besonders Marianne von Werefkin Kandinsky verteidigte, zogen sie und Jawlensky nach der Ausjurierung des Bildes keine Konsequenzen, sondern blieben Mitglieder der Vereinigung – allerdings auch nur für ein weiteres Jahr.

Typisch für alle Künstlervereinigungen der Zeit war ihre Kurzlebigkeit. Überall schlossen sich Künstler gegen die alte akademische Kunst zusammen, ohne eine gemeinsame Plattform zu besitzen. Ihre unterschiedlichen Kunstauffassungen machten sich mit der Zeit bemerkbar und führten oft zu unerfreulichen Auseinandersetzungen, die über kurz oder lang aus nichtigen Anlässen zum Austritt von einer der beiden Seiten führten. So hatte es auch in der Wiener Secession – allerdings erst nach acht Jahren – einen Streit gegeben, in dessen Verlauf diejenigen Künstler, die maßgeblich am Aufbau der Secession beteiligt gewesen waren, ihren Auszug erklären mußten, da sie in der Kampfabstimmung unterlagen. Ähnliches spielte sich jetzt in München ab, mit dem Unterschied, daß die Neue Künstlervereinigung gerade drei Jahre bestand und Franz Marc nur zehn Monate Mitglied gewesen war.

ABBILDUNG SEITE 34/35:
Die kleinen gelben Pferde, 1912
Öl auf Leinwand, 66 × 104,5 cm
Stuttgart, Staatsgalerie Stuttgart

Das Bild des Tieres – Fremdheit oder Vermenschlichung

Am 3. Juli 1989 stand in der Süddeutschen Zeitung: »Den höchsten Zuschlag, den jemals ein Kunstwerk auf einer deutschen Auktion erreichte, bekam bei der 9. Versteigerung der ›Villa Griesebach‹ in Berlin die Tempera *Fabeltiere* von Franz Marc. Das 1913 gemalte, nur 26,5 × 30,5 Zentimeter große, mit Tusche monogrammierte Bild erzielte netto 2600000 Mark... Die Taxe... hatte maximal 600000 Mark betragen.« (Abb. S. 74) Noch im September 1980 hatte die Zeitschrift »art« erklärt, daß der »Marc-Boom« abgeebbt sei, der in den fünfziger und sechziger Jahren geherrscht habe.

Es ist richtig, daß die Reproduktionen, die damals Wohnzimmer und Schulbücher gleichermaßen schmückten, inzwischen verschwunden sind. Das Auktionsergebnis zeigt jedoch, daß Franz Marc immer noch hoch im Kurs steht.

Die blauen Pferde, die weißen und gelben Kühe, die roten Rehe rühren den Betrachter immer noch an, er erfährt keine Distanz zu den Bildern, sondern wird in ein Miterleben mit den Protagonisten, den Tieren, hineingezogen. Das ist wahrscheinlich auch ein Grund, weshalb diese Bilder nach 1945 derart populär wurden. Der immer wieder beschworene Nachholbedarf, durch die moderne-feindliche Zeit des tausendjährigen Reiches verursacht, erstreckte sich ja nicht auf das gesamte Spektrum der modernen Kunst, bezog nicht alle Künstler ein, die spätestens seit der berühmten Ausstellung von 1937 als entartet galten. Abstrakte, dadaistische oder gar sozialkritische Bilder waren noch immer nicht gefragt. Das »Nachholbedürfnis« beschränkte sich auf blaue Pferde und Sonnenblumen und damit im Grunde auf eine Flucht aus der Wirklichkeit. Den bevorzugten Malern wurde man damit häufig nicht gerecht. So fanden sich in den Schulbüchern – oder im Angebot an Reproduktionen – vom Meister des *Turms der blauen Pferde* keine abstrakten Bilder, sondern eben die blauen Pferde und roten Rehe.

Auch wenn das Werk von Franz Marc viele andere Momente enthält, so spielt die Flucht aus der Wirklichkeit doch gerade in seinen Tierbildern eine große Rolle. Daß der Betrachter keinen Abstand erfährt, daß er sich hineinfühlt und berührt wird, erreichte Marc mit bestimmten, ihm zum Teil gar nicht bewußten Stilmitteln: Er selber hatte immer wieder den Anspruch vertreten, das Tier nicht aus dem Blickwinkel des Menschen, sondern aus dem des Tieres zu malen. Den ersten Anstoß dazu gab dem Tierliebhaber, der schon als Junge immer mit seinem Hund unterwegs war, der Tiermaler Jean Bloé Niestlé, der später mit Marc und dem rheinischen Maler Heinrich Campendonk (1889–1957)

Springende Pferdchen, 1912
Holzschnitt, 13,2 × 9 cm
München, Städtische Galerie im Lenbachhaus

Reh im Walde II, 1912
Öl auf Leinwand, 110,5 × 80,5 cm
München, Städtische Galerie im Lenbachhaus

zusammen in Sindelsdorf in einer Art kleiner Künstlerkolonie lebte. Er hatte, wie wir uns erinnern, Marc auf den Gedanken gebracht, sich mit künstlerischer Absicht in die »Seele« des Tieres zu versetzen.

Ab 1910 hat Franz Marc in seinen veröffentlichten Schriften und persönlichen Aufzeichnungen immer wieder auf das Tier und die Kunst Bezug genommen. Seine Gedanken formulierte er von Jahr zu Jahr deutlicher. 1910 verfaßte er auf Bitte des Verlegers Reinhard Piper einen kurzen Text »Über das Tier in der Kunst«. Er wurde in dem von Piper selbst geschriebenen Buch »Das Tier in der Kunst« neben der von Marc 1908/09 geschaffenen bronzenen Pferdegruppe veröffentlicht. Marc distanzierte sich darin als erstes von der herkömmlichen Tiermalerei: »Meine Ziele liegen nicht in der Linie besonderer Tiermalerei. Ich suche einen guten, reinen und lichten Stil, in dem wenigstens ein Teil dessen, was mir moderne Maler zu sagen haben werden, restlos aufgehen kann. Ich... suche mich pantheistisch einzufühlen in das Zittern und Rinnen des Blutes in der Natur, in den Bäumen, in den Tieren, in der Luft...« Er kritisierte das alte Staffeleibild und erkannte dessen Überwindung bei den Franzosen, von Delacroix angefangen bis zu den Pointillisten. Verwundert stellte er allerdings fest, daß sie demjenigen Thema, das für ihn auf der Hand lag, in ihrer Malerei systematisch aus dem Weg gegangen seien, nämlich dem Tierbild: »Ich sehe kein glücklicheres Mittel zur ›Animalisierung der Kunst‹ als das Tierbild. Darum greife ich danach. Bei einem van Gogh oder einem Signac ist alles animalisch geworden, die Luft, selbst der Kahn, der auf dem Wasser ruht, und vor allem die Malerei selbst. Diese Bilder haben gar keine Ähnlichkeit mehr mit dem, was man früher ›Bilder‹ nannte.« Seine Plastik der Pferdegruppe war für ihn »ein tastender Versuch nach derselben Richtung«. Er wollte darin dem Betrachter die Möglichkeit nehmen, nach dem Pferdetyp zu fragen, er sollte vielmehr das »innerlich zitternde Tierleben herausfühlen«.[11]

Eineinhalb Jahre später formulierte der Künstler – allerdings in einer nicht zur Veröffentlichung bestimmten Aufzeichnung – sehr viel radikaler: »Wie sieht ein Pferd die Welt oder einen Adler, ein Reh oder ein Hund? Wie armselig, seelenlos ist unsere Konvention, Tiere in eine Landschaft zu setzen, die unseren Augen zugehört statt uns in die Seele des Tieres zu versenken, um dessen Bilderkreis zu erraten.« Und weiter: »Was hat das Reh mit dem Weltbild zu thun, das wir sehen? Hat es

Eselfries
ägyptisch, um 2700–2600
Leiden, Rijksmuseum

Eselfries, 1911
Öl auf Leinwand, 81 × 150 cm
Privatbesitz

irgendwelchen vernünftigen oder gar künstlerischen Sinn, das Reh zu malen, wie es unsrer Netzhaut erscheint oder in kubistischer Form, weil wir die Welt kubistisch fühlen? Wer sagt mir, daß das Reh die Welt kubistisch fühlt; es fühlt sie als ›Reh‹, die Landschaft muß also »Reh« sein... Ich kann ein Bild malen: das Reh. Pisanello hat solche gemalt. Ich kann aber auch ein Bild malen wollen: ›das Reh fühlt‹. Wie unendlich feinere Sinne muß ein Maler haben, das zu malen!«[12]

1912/13 hielt er – ebenfalls in seinen privaten Aufzeichnungen – einige Thesen zur abstrakten Kunst und zu den Grenzen der Kunst fest. Auch hier bemühte er wieder das Tier: »Wir werden nicht mehr den Wald oder das Pferd malen, wie sie uns gefallen oder scheinen, sondern *wie sie wirklich sind*, wie sich der Wald oder das Pferd selbst fühlen, ihr absolutes Wesen, das hinter dem Schein lebt, den wir nur sehen.« Aus dieser Vision folgerte er: »Wir müssen von nun an verlernen, die Tiere und Pflanzen auf uns zu beziehen und unsre Beziehung zu ihnen in der Kunst darzustellen. Das ist vorbei, muß vorbei sein oder wird eines Tages, – oh der glückliche Tag! – vorbei sein.«[13]

Im Krieg faßte er seine Gedanken über die Kunst in einem Brief an Maria Marc noch einmal zusammen. Am 12. April 1915 schrieb er an sie: »Ich denke viel über meine eigene Kunst nach. Der Instinkt hat mich im Großen und Ganzen, auch bisher nicht schlecht geleitet, wenn die Werke auch unrein waren; vor allem der Instinkt, der mich von dem Lebensgefühl für den Menschen zu dem Gefühl für das animalische, den ›reinen Tieren‹ wegleitete. Der unfromme Mensch, der mich umgab, (vor allem der männliche) erregte meine wahren Gefühle nicht, während das unberührte Lebensgefühl des Tieres alles Gute in mir erklingen ließ... Ich empfand schon sehr früh den Menschen als ›häßlich‹; das Tier schien mir schöner, reiner.«[14]

»Wir erleben heute eine der bedeutendsten Momente in der Kulturgeschichte. Alles was wir von ›alter‹ Kultur (Religion, Monarchentum, Adel, Privilegien, (auch rein geistige), Humanismus etc.) noch mit uns schleppen, ist eine ›Gegenwart, die schon der Vergangenheit angehört‹... welcher Art neuer Kultur wir entgegengehen, wird kaum heute jemand sagen können, weil wir eben selber mitten in der Wandlung stehen; wir modernen Maler sind kräftig mit am Werk, für das kommende Zeitalter, das alle Begriffe und Gesetze neu aus sich gebären wird, auch eine ›neugeborene‹ Kunst zu schaffen; sie muß so rein und kühn werden, daß sie ›alle Möglichkeiten‹ zuläßt, die die neue Zeit an sie stellen wird.« *Franz Marc, 21.1.1911*

Die gelbe Kuh, 1911
Öl auf Leinwand, 140 × 190 cm
New York, Solomon R. Guggenheim Museum

Marcs intensive Auseinandersetzung mit der Darstellung des Tieres war in seiner Gefühlsbindung an Tiere begründet. Sein Russi, ein großer sibirischer Schäferhund, begleitete ihn fast überallhin. Er findet sich auf vielen Bildern wieder. Katzen lebten spätestens seit dem Umzug nach Sindelsdorf bei den Marcs, und später erfüllte Franz Marc sich den lang gehegten Wunsch nach zwei Rehen namens Schlick und Hanni.

Den Anspruch, sich in das Tier hineinzuversetzen, es nicht mit den Augen des Menschen, sondern aus seiner eigenen Sicht zu malen, versuchte er auch durch eingehende Beobachtungen auf den Pferdekoppeln und Kuhweiden zu verwirklichen. Wenn er in Berlin bei den Schwiegereltern war, führte ihn der Weg immer wieder in den Zoo.

Die Beschäftigung mit der modernen Malerei und der zu Beginn des Jahrhunderts überall und immer wieder formulierte Anspruch, die Dinge nicht so zu malen, wie wir sie sehen, sondern wie sie wirklich sind, hat wohl seine Bestrebungen angeregt, die Tiere nicht im herkömmlichen Sinne zu malen, sondern ihr Fühlen einzufangen. Er hat dabei allerdings unser aller Neigung außer acht gelassen, das Tier zu vermenschlichen, die sicherlich aus der Unmöglichkeit herrührt, sich in ein artfremdes Lebewesen hineinzuversetzen. Zeugnisse von Zeitgenos-

Das Äffchen, 1912
Öl auf Leinwand, 70,4 × 100 cm
München, Städtische Galerie im Lenbachhaus

sen belegen diese Vermenschlichung des Tieres durch Marc. So schrieb Wassily Kandinsky in seinen Erinnerungen an den Maler: »Dort zeigte mir Marc seine Rehe, die er wie eigne Kinder liebte ...« Die eigentlich für das malerische Werk bedeutungslose Tatsache, daß Franz und Maria Marc die Erfüllung ihres Kinderwunsches, der durch den Briefwechsel zwischen Maria Marc und Lisbeth Macke belegt ist, versagt blieb, gewinnt hier an Bedeutung. Die Tiere waren seine ihm verwehrten Kinder. Paul Klee formulierte das Verhalten den Tieren gegenüber sehr viel allgemeiner und doch präziser, als er nach Marcs Tod in sein Tagebuch notierte: »Menschlicher ist er, er liebt wärmer, ausgesprochener. Zu den Tieren neigt er sich menschlich. Er erhöht sie zu sich.«

Klee hatte mit diesen Sätzen nicht beabsichtigt, das malerische Werk von Franz Marc zu interpretieren. Es war vielmehr ein Vergleich zwischen dem verstorbenen Freund und ihm selbst, der ihn zu dieser Tagebuchnotiz bewogen hatte. Trotzdem – oder gerade deshalb – sagen diese wenigen spontan hingeschriebenen Sätze sehr viel über das Tier in den Gemälden Marcs aus. Vor kurzem wurde aus kunsthistorischer Sicht glaubhaft nachgewiesen, daß Marc bei seinen Tierbildern auf typische Figurenkompositionen zurückgegriffen hat, auch wenn dem Maler dies wahrscheinlich nicht bewußt war. Diese Tatsache erklärt, warum es schwierig ist, diese Bilder mit Distanz zu betrachten. Marcs Bestreben, sich in die Tiere hineinzuversetzen, führte dazu, daß er sie vermenschlichte und in den meisten Bildern menschliche Gefühle, wenn nicht gar sich selbst darstellte.

Wenn man die Bilder aus diesem Blickwinkel betrachtet, muß man die Farbensymbolik berücksichtigen, die Marc für seine Kompositionen entwickelt hatte. Das Gemälde *Blaues Pferd I* von 1911 (Abb. S. 18) wird dann, genauso wie der *Tote Spatz* von 1905 (Abb. S. 11), zu einer Selbstdarstellung des Malers. »Blau ist das männliche Prinzip, herb und geistig«, hatte Marc an Macke geschrieben. Pferde waren für Marc offensichtlich häufig mit dem männlichen Prinzip verbunden. Das Pferd, welches den Vordergrund des Bildes beherrscht, steht fest auf seinen vier Beinen. Den Kopf hat es etwas zur Seite geneigt und blickt nachdenklich nach unten. Die hügelige Landschaft ist sehr stark abstrahiert. Allein eine stilisierte Pflanze im Vordergrund, deren mittleres Blatt steil emporwächst, deutet auf lebende Natur hin, wie sie unseren Wahrnehmungsgewohnheiten entspricht. In dieser Landschaft sind sämtliche Farben des Farbkreises, also Gelb, Grün, Blau, Rot, ein leichtes Violett und Orange, vertreten. Sie unterstützen das Blau des Pferdes, bilden einen harmonischen Farbklang und bewirken, daß von dem stehenden

Rote Rehe II, 1912
Öl auf Leinwand, 70 × 100 cm
München, Staatsgalerie moderner Kunst

Affenfries, 1911
Öl auf Leinwand, 76 × 134,5 cm
Hamburg, Hamburger Kunsthalle

Pferd eine ungeheure Ruhe ausgeht. In sich zentriert – es blickt nicht aus dem Bild, sondern nach unten –, strahlt es Selbstbewußtsein aus. Marc hat dieses Bild gemalt, nachdem er bereits in die Künstlervereinigung gewählt worden war. Er hatte mit seinen Bildern endlich Erfolg und fühlte sich stark genug, um für die neue Kunst zu kämpfen, wie die Antwort auf den Vinnen-Protest deutlich gemacht hat. In diesem Sinne kann das Gemälde *Blaues Pferd I* als eine Art Selbstporträt verstanden werden. Das Bild *Die gelbe Kuh* (Abb. S. 40) drückt Lebensfreude aus, mitgeteilt durch »Gelb, das weibliche Prinzip, sanft, heiter und sinnlich«. In einer sehr viel konkreter dargestellten Landschaft, die in grünen, roten, gelblichen und schwarz-grauen Farben gehalten ist, springt eine massige, schwerfällige Kuh durch das Bild, den Kopf genießerisch erhoben. Die Vorderbeine stemmen sich unter der Körperlast auf den Boden, die Hinterbeine befinden sich noch im Sprung. Diese Bewegtheit – man hat den Eindruck, daß das Tier gleich aus dem Bildraum entschwindet – steht im Kontrast zu der ruhigen Landschaft, wo im Hintergrund friedlich Tiere grasen.

Franz Marc beschäftigte sich 1911 intensiv mit der Kunst der außereuropäischen Völker, wie so viele Malerkollegen in dieser Zeit. Durch einen Brief an Macke ist belegt, daß er im Januar 1911 häufig das Völkerkundemuseum in Berlin besuchte. Er wird auch Abbildungen aus den damals immer häufiger werdenden Publikationen gekannt haben. In einem Fall, dem *Eselsfries* (Abb. S. 39), sind die Vorbilder, ein altägyptischer Eselsfries (Abb. S. 38) und Tiermasken aus Kamerun, bekannt. Auch die *Gelbe Kuh* wurde mit einem mykenischen Goldbecher, auf dem Stierfangszenen dargestellt sind, in Verbindung gebracht: Auf dem

Tiger, 1912
Holzschnitt, 20 × 24 cm

Becher ist ein Tier in der gleichen springenden Haltung dargestellt. Marc weitete jedoch die Beschäftigung mit dem Primitivismus nicht aus, wie das beispielsweise die Maler der Künstlergruppe »Die Brücke« getan haben. Es ist auch zu bezweifeln, daß das Gelb der Kuh auf das Gold des Bechers zurückzuführen ist. Hier spielt eher die Farbensymbolik, die Marc für sich entwickelt hatte, die entscheidende Rolle.

Sein System der Farbzuordnungen war jedoch nicht dogmatisch, wie das berühmte Bild *Tiger* (Abb. S. 51) von 1912 zeigt, wo das Gelb weder als weiblich noch als sanft, heiter und sinnlich aufzufassen ist. Die weiche, hügelige Landschaft wurde hier von scharfkantigen, bizarren, eckigen Formen abgelöst, die auf den Einfluß des Kubismus verweisen mögen (was immer wieder behauptet wird), vor allem aber inhaltlich zu erklären sind. Inmitten dieser grünen, violetten, roten und blauen Formen liegt der gelb-schwarze Tiger, zum Sprung bereit. Er wendet dem Betrachter den Rücken zu, hat den Kopf gedreht und blickt seitlich aus dem Bild heraus. In diesem Blick spürt man Bedrohung, Stärke des Tieres, Gefahr, was durch die scharfkantigen Formen betont wird. Hat Marc hier wirklich die Gefühle des Tigers ausgedrückt, oder hat er nicht eher menschliche Befindlichkeiten auf den Tiger übertragen?

»...es fühlt sie als ›Reh‹, die Landschaft muß also ›Reh‹ sein«, schrieb Marc im Winter 1911/12. Betrachtet man das Bild *Reh im Walde II* (Abb. S. 36) von 1912, sieht man ein kleines gelbes Reh in einem Meer von Farbformen liegen, ein Baum im Vordergrund durchschneidet die Fläche diagonal. Der vordere Teil des Bildes, in welchem sich das Reh zusammengekauert hat, wirkt ruhig und beschützend, während im Hintergrund die Formen in Bewegung geraten und den Eindruck erwecken, als gehe ein Sturm durch den Wald. Spontan versetzt sich der Betrachter

Der Turm der blauen Pferde, 1913
Öl auf Leinwand, 200 × 130 cm
verschollen

Zwei Katzen, blau und gelb, 1912
Öl auf Leinwand, 74 × 98 cm
Basel, Öffentliche Kunstsammlung,
Kunstmuseum Basel

in das Reh, welches geborgen ist, sich in Sicherheit befindet. Der Anspruch, »Reh« zu fühlen, kann nicht eingelöst werden. Vielmehr ist im Reh »Mensch« enthalten.

In seinen Tierdarstellungen hat Marc den objektiven Standpunkt des Betrachters verlassen. Er war nicht darum bemüht, Bewegungen der Tiere um der Bewegungen willen einzufangen. In seinem Versuch, sich in die Tiere hineinzuversetzen, von ihnen aus zu sehen, zu malen, malte er das Menschliche. »Er erhöht sie zu sich«, wie Paul Klee richtig formulierte.

Die Erhöhung des Tieres, des Pferdes, ist am besten im berühmten *Turm der blauen Pferde* (Abb. S. 45) nachzuvollziehen. Da das Bild seit 1945 verschollen ist, sind wir heute auf Reproduktionen angewiesen, die uns die Größe und damit den tatsächlichen Eindruck, den das Original auf den Betrachter machte, nicht mehr vermitteln können. Nur Augenzeugenberichte können uns die Ausstrahlung des Bildes heute noch vergegenwärtigen: »Eine Leinwand von 2 Meter Höhe und 1,30 Meter Breite schlägt uns in ihren Bann… Wie eine Vision leuchtet dicht vor uns eine Gruppe von vier Pferden auf… Der mächtige Körper des vor-

Blauschwarzer Fuchs, 1911
Öl auf Leinwand, 50 × 63,5 cm
Wuppertal, Von der Heydt-Museum

deren Tieres mißt nur wenig unter Lebensgröße. Das Pferd scheint aus der Tiefe nach vorn zu drängen und unmittelbar vor dem Beschauer zu verhalten...«[15]

Die vier frontal übereinandergestaffelten Tiere sind aus der Mittelachse nach rechts gerückt. Die Mächtigkeit der Leiber wird durch den sich viermal wiederholenden Bewegungsablauf unterstrichen. Sie stehen so nach rechts gewandt, daß ihre Kruppen die Mitte der Komposition bilden, drehen aber alle den Kopf nach links ins Bild, wo eine sehr stark abstrahierte Felsenlandschaft in den Bildraum hineinragt. Ein orangefarbener Regenbogen, der von einem gelben Farbfeld hinterfangen wird, schließt das Bild nach oben hin ab. Dieser Regenbogen, wie auch die Mondsichel auf der Brust des ersten Pferdes und die durch Kreuze angedeuteten Sterne auf seinem Leib lassen die Vermutung zu, daß Marc hier die Einheit von Kreatur und Kosmos darstellen wollte. Der Mensch, der in dieser Verschmelzung von Natur, Tier und Kosmos nicht vorkommt, ist in der Überhöhung der Pferde präsent. Sie demonstrieren eine Macht, die dem Menschen eigen ist. Darin liegt auch die Faszination, die von diesem Werk ausgeht.

Kühe, gelb-rot-grün, 1912
Öl auf Leinwand, 62 × 87,5 cm
München, Städtische Galerie im Lenbachhaus

In vielen der späteren Bilder lösen sich die Tiere immer mehr in formelhafte Gestalten auf, was noch im Rahmen der Stilentwicklung Thema sein wird. Im Zusammenhang mit der Frage, wie weit die Tiere überhaupt um ihrer selbst willen dargestellt werden, ist das späte Bild *Vögel* (Abb. S. 75) von 1914 wichtig. Aus farbigen Formen, die alle an Flügel erinnern, schälen sich mehrere Vögel heraus. Das ganze Gemälde scheint zu fliegen. Vor kurzem ist auf einen Text verwiesen worden, dem dieses Bild inhaltlich zugeordnet werden kann und der zeigt, wie sehr Marc sein Gedankengut in seine Bilder einarbeitete und damit sich selbst in die Tiere hineinprojizierte. Seinen programmatischen Aufsatz »Die neue Malerei« von 1912, der Beckmann zum Widerspruch reizte, leitete er mit dem folgenden Satz ein: »Die denkwürdigsten Jahre der modernen Kunstentwicklung bleiben die 90er Jahre des vergangenen Jahrhunderts, in denen der französische Impressionismus sich in seinem eigenen Feuer verzehrte, während aus seiner Asche sich phönixgleich ein Schwarm neuer Ideen erhob, Vögel mit bunten Federn und mystischen Schnäbeln.«[16] Der Aufbruch zu neuen Ufern, von dem Marc in diesem Satz spricht, kommt auch in dem Bild zeichenhaft zur Darstel-

Das kleine blaue Pferdchen, 1912
Öl auf Leinwand, 57,5 × 73 cm
Saarbrücken, Saarland-Museum

lung; besonders in dem fliegenden Vogel in der Mitte und in den aufwärtsstrebenden Formen.

1914 hatte Marc begonnen, auch ungegenständliche Bilder zu malen. Das Tier hatte für ihn die Bedeutung verloren, die es vordem gehabt hatte. In dem oben zitierten Brief an Maria Marc vom 12. April 1915, der als Zusammenfassung seines künstlerischen Schaffens gelten kann, schrieb er: »Und vom Tier weg leitete mich ein Instinkt zum Abstrakten, das mich noch mehr erregte;... Ich empfand schon sehr früh den Menschen als ›häßlich‹; das Tier schien mir schöner, reiner; aber auch an ihm entdeckte ich soviel gefühlswidriges u. häßliches, sodaß meine Darstellungen instinktiv... immer schematischer, abstrakter wurden.«[17]

Marc bediente sich jetzt nicht mehr des Tieres, um sich und seine Gefühle, seine Innerlichkeit, darzustellen. Er löste sich immer weiter vom Gegenstand, bis die ersten abstrakten Werke entstanden. An diesem Punkt brach seine Entwicklung ab.

Träumendes Pferd, 1913
Wasserfarben auf Papier, 39,4 × 46,3 cm
New York, Solomon R. Guggenheim Museum

Tiger, 1912
Öl auf Leinwand, 111 × 111,5 cm
München, Städtische Galerie im Lenbachhaus

Vom »Blauen Reiter« zum »Ersten Deutschen Herbstsalon«

Franz Marc und der »Blaue Reiter« sind voneinander nicht zu trennen. Zahlreiche Publikationen geben über die Geschichte des »Blauen Reiters« Auskunft. Hier soll nur die Rolle, die Marc gespielt hat, beleuchtet werden, denn sie war prägend für sein späteres Handeln und Arbeiten.

Wenn heute vom »Blauen Reiter« die Rede ist, denkt man an eine Künstlergruppe in Schwabing, angeführt von Wassily Kandinsky und Franz Marc. Diese Vorstellung täuscht. Kandinsky hatte, nachdem er das Manuskript der Schrift »Über das Geistige in der Kunst« abgeschlossen hatte, den Plan, eine Art Almanach herauszugeben, in dem nur Künstler – aber nicht nur bildende Künstler – zu Wort kommen sollten. Allein sah er sich zur Verwirklichung einer solchen Idee nicht imstande.

»Und da kam Franz Marc aus Sindelsdorf. Eine Unterredung genügte: wir verstanden uns vollkommen.« So einfach, wie Kandinsky dies im Rückblick 1930 beschrieb, hatte sich der Beginn der Zusammenarbeit jedoch nicht zugetragen. Es ist zu vermuten, daß Kandinskys Wohlwollen dem vergleichsweise jungen Malerkollegen gegenüber genährt wurde, als sie unabhängig voneinander den Plan faßten, der Kampfschrift Vinnens einen Protest entgegenzusetzen und diesen dann gemeinsam durchführen konnten. Kandinsky lernte im Zuge dieser Unternehmung die Fähigkeiten des neu gewonnenen Freundes bei der Herausgabe von Schriften kennen. Außerdem hatten er und Marc ähnliche Auffassungen über Malerei, was sich bereits in Marcs zustimmender Kritik über die zweite Ausstellung der Künstlervereinigung gezeigt hatte, in der er den Begriff des »Geistigen« ähnlich verwendet hatte wie Kandinsky.

Schlafende Hirtin, 1912
Holzschnitt, 19,8 × 24,1 cm

Marc selber hatte bereits von der Herausgabe einer Zeitschrift geträumt. Er gehörte nicht zu denen, die sagen, »Maler male, rede nicht«, sondern wollte die Bilder, die er malte, die Kunst, die er und seine Freunde schufen, auch durch das geschriebene Wort und durch die Präsentation erläutern.

Am 19. Juni 1911 schrieb Kandinsky an Marc den zur Legende gewordenen Brief, in dem er dem jungen Freund vorschlug, gemeinsam eine Art Almanach herauszugeben. Dort sollten Bilder verschiedener Kulturkreise einander gegenübergestellt, dazu Schriften von Künstlern veröffentlicht werden. Eine Antwort von Franz Marc auf den Brief ist nicht erhalten. Da er aber häufig mit Kandinsky zusammentraf, ist es fraglich, ob es jemals eine schriftliche Antwort gegeben hat.

Die Arbeiten an dem Almanach begannen sehr bald nach dem Brief. Im September weihte Marc August Macke in die Pläne ein und bat ihn

Die verzauberte Mühle, 1913
Öl auf Leinwand, 130,6 × 90,8 cm
Chicago, The Art Institute of Chicago

Die kleinen blauen Pferde, 1911
Öl auf Leinwand, 61 × 101 cm
Stuttgart, Staatsgalerie Stuttgart,
Sammlung Lütze

»... was über mich mögen sich die Menschen alles dabei denken, wenn sie sie [meine Bilder] sehen! Es quält mich, daß keines so klar ist, daß man meinen Wunsch unzweideutig lesen kann, den Wunsch zur Religion, die nicht da ist; aber man kann deswegen doch nicht das Malen aufstecken, weil man um 50 oder 100 Jahre zu früh auf diesen Planeten geraten ist. Wenn man das könnte: Den Kopf unter die Decke stecken, für 100 Jahre, und dann von vorn anfangen.« *Franz Marc, 31. 7. 1912*

mitzuarbeiten. Der kam im Oktober und genoß offensichtlich die arbeitsreiche Atmosphäre, die in Sindelsdorf und Murnau herrschte, wo man sich abwechselnd traf. Über den Namen war man sich im September einig geworden. In seinen Erinnerungen beschreibt Kandinsky sehr einfach: »Den Namen ›Der blaue Reiter‹ erfanden wir in der Gartenlaube in Sindelsdorf; beide liebten wir Blau, Marc – Pferde, ich – Reiter. So kam der Name von selbst.«[18]

Während der Arbeit an dem Almanach wuchsen die Spannungen in der Neuen Künstlervereinigung. Noch bevor der Almanach bei Piper erscheinen konnte, kam es zu der denkwürdigen Sitzung, in der Marc und Kandinsky ihren Austritt erklärten. Innerhalb von vierzehn Tagen gelang es ihnen, neben der Ausstellung der Künstlervereinigung in den anschließenden Räumen der Galerie Thannhauser eine Ausstellung auf die Beine zu stellen, in der 50 Werke von 14 Künstlern zu sehen waren. Wie auch im Almanach kam es ihnen darauf an, sehr verschiedene Erscheinungen in der neuen Malerei auf internationaler Basis nebeneinanderzustellen.

Diese Ausstellung, in der Gemälde von Albert Bloch, David und Wladimir Burljuk, Heinrich Campendonk, Robert Delaunay, Elisabeth Epstein, Eugen von Kahler, Wassily Kandinsky, August Macke, Franz Marc, Gabriele Münter, Jean Bloé Niestlé, Henri Rousseau und dem Komponisten Arnold Schönberg zu sehen waren, reiste in den nächsten Jahren nach Köln, Berlin, Bremen, Hagen, Frankfurt, Hamburg, Rotterdam, Amsterdam, Barmen, Wien, Prag, Budapest, Königsberg, Oslo, Lund, Helsingfors, Stockholm, Trondheim und Göteborg. In Berlin war sie in der neu eröffneten, von Herwarth Walden geleiteten Galerie »Der Sturm« zu sehen. Walden war es auch, der sich vor allem für die Stationen in den nordischen Ländern stark machte.

Die Ausstellung wurde von den beiden »Machern« ebenfalls »Der Blaue Reiter« genannt. Kurz darauf zeigten sie in der Kunsthandlung Goltz Druckgraphik und Aquarelle. In dieser zweiten und letzten Ausstellung der Redaktion des »Blauen Reiters« waren die Franzosen sehr viel stärker vertreten. Georges Braques, Roger de la Fresnaye, André Derain, Pablo Picasso, Robert Lotrion und Maurice Vlaminck zeigten neben Robert Delaunay ihre Werke.

Marc hatte an Weihnachten 1911 in Berlin eine bis dahin in München völlig unbekannte Künstlergruppe kennengelernt, die sich »Die Brücke« nannte, und hatte Druckgraphiken von Erich Heckel, Ernst Ludwig Kirchner, Otto Mueller und Max Pechstein für die Ausstellung mitgebracht. Es war die erste Berührung dieser beiden so wichtigen expressionistischen Strömungen innerhalb Deutschlands, die in der Literatur häufig als Konkurrenzunternehmen dargestellt worden sind, obwohl sie sich selber nicht so verstanden.

Neben anderen deutschen und auch Schweizer Künstlern war auch Paul Klee vertreten. Von den Russen schickten, vermittelt durch Kandinsky, Natalia Gontscharowa, Michail Larionow und Kasimir Malewitsch Werke für die Ausstellung. Bemerkenswert und bis heute wichtig ist die Internationalität, die sich in diesen Ausstellungen manifestierte und die so ganz im Gegensatz zu den nationalistischen Tendenzen stand, wie sie zum Beispiel im Vinnen-Protest zu Tage getreten waren. Die erste Ausstellung der Redaktion des »Blauen Reiters« wird heute als die Geburtsstunde der Moderne in Deutschland bezeichnet.

Im Mai 1912 erschien dann endlich nach vielen Mühen und Streitereien mit Reinhard Piper, der selbstverständlich wieder als Verleger fungierte, nach wiederholten Änderungen des Inhaltsverzeichnisses, weil die Beiträge entweder den Vorstellungen der Herausgeber zuwider liefen oder nicht rechtzeitig abgegeben wurden, der Almanach, der keiner mehr war. Piper hatte kurz vor der Drucklegung darauf bestanden, daß dieses Wort aus dem Titel gestrichen werden müsse. Kandinsky hatte daraufhin dieses Wort aus seinem Holzschnitt noch herausgeschnitten. Auch die beiden Herausgeber sprachen inzwischen nicht mehr von einem jährlich erscheinenden Organ, sondern von einer in lockerer Folge erscheinenden Reihe. Dieser Plan wurde jedoch ebensowenig realisiert. Es blieb beim ersten Band des »Blauen Reiters«.

Hundertvierzig Abbildungen standen neunzehn Artikeln gegenüber. Neben Bildern der Moderne wurden völkerkundliche Objekte, mittelalterliche Kunstwerke und Volkskunst, insbesondere Hinterglasbilder aus Bayern gezeigt. Für Marc und Kandinsky lag die Parallele in der Unmittelbarkeit, mit der all diese Kunstwerke geschaffen worden waren.

Drei kurze Texte von Marc leiteten den Band ein, drei lange von Kandinsky beschlossen ihn. Dazwischen kamen bildende Künstler wie August Macke oder David Burljuk, aber auch Musiker wie Arnold Schönberg und der Russe Thomas von Hartmann zu Wort.

»Der Blaue Reiter« faßte nicht Gewachsenes zusammen, er bot keine systematischen Zusammenfassungen der damaligen Kunsttheorien. Vielmehr erschien er am Beginn einer Entwicklung, die 1914 erst einmal zu einem Ende kam: die Entwicklung der Malerei hin zum abstrakten Bildwerk. Der Almanach, wie er einfachheitshalber immer noch genannt wird, kann als ein Programm der modernen Ästhetik angesehen werden, in dem, wenn auch unvollständig, Grundsätze bildnerischen Schaffens angesprochen wurden.

Kandinsky versuchte in dem Aufsatz »Über die Formfrage« den Begriff der »inneren Notwendigkeit«, den er in »Über das Geistige in der Kunst« bereits eingeführt hatte, noch weiter zu konkretisieren. Hier – und nicht, wie oft irrtümlich angenommen, in der vorher erschienenen Schrift – führte er die Begriffe der »großen Abstraktion« und der »großen Realistik« ein, die heute noch ihre Gültigkeit besitzen. Werner Hofmann hat in seinem wichtigen Werk über die »Grundlagen der modernen Kunst« vorgeführt, daß dieser Aufsatz und vor allem die beiden Begriffe für die Künstler bis in die heutige Zeit von entscheidender Bedeutung sind. Wenn auch Marc nicht als einer der Hauptväter der Moderne gilt, so hat ihm Kandinsky in dem Aufsatz diesen Stellenwert doch zuerkannt: »Das starke abstrakte Klingen der körperlichen Form verlangt nicht durchaus Zerstörung des Gegenständlichen. Daß es auch hier keine allgemeine Regel gibt, sehen wir im Bilde von Marc (*Der Stier*). Es kann also der Gegenstand den inneren und den äußeren Klang vollkommen behalten, und dabei können seine einzelnen Teile zu selbständig klingenden abstrakten Formen sich verwandeln und also einen gesamten abstrakten Hauptklang verursachen.«[19]

Franz Marc selbst hat in seinen kurzen Artikeln für den »Almanach« Thesen aufgestellt, die bis heute nichts von ihrer Aktualität verloren haben.

In dem ersten Text »Geistige Güter« stellte er fest, daß sich die Menschheit zwar über die Eroberung neuer Kolonien oder über neue technische Errungenschaften freue, aber nie über neue geistige Güter,

Versöhnung, 1912
Zu dem gleichnamigen Gedicht von Else Lasker-Schüler
Holzschnitt, 20 × 25,8 cm

Im Regen, 1912
Öl auf Leinwand, 81,5 × 106 cm
München, Städtische Galerie im Lenbachhaus

womit er besonders die neue Kunst im Auge hatte. Er bedauerte »die allgemeine Interesselosigkeit der Menschen für neue geistige Güter« und formulierte zum Schluß den heute noch gültigen Satz: »Neue Ideen sind nur durch ihre Ungewohnheit schwerverständlich...«

In dem zweiten Text beschrieb er »Die ›Wilden‹ Deutschlands«. Sie würden gegen eine »alte, organisierte Macht« kämpfen, die ihnen von der Anzahl her zwar überlegen sei, doch in einem solchen Fall siege immer die »Stärke der Ideen«. Zu diesen »Wilden« zählte er die »Brücke«, die »Neue Sezession« in Berlin und die »Neue Künstlervereinigung« (offensichtlich hatte er den Text vor seinem Austritt geschrieben). Nachdem er kurz die Entstehungsgeschichte der drei Gruppen beschrieben hatte, wehrte er sich gegen die immer wieder aufgestellte Behauptung, daß sich diese Kunst aus dem Impressionismus entwickelt habe. Das Ziel dieser »Wilden« erklärte er in den immer wieder zitierten und für uns heute sehr pathetisch klingenden Worten: »Durch ihre Arbeit ihrer Zeit Symbole zu schaffen, die auf die Altäre der kommenden geistigen Religion gehören und hinter denen der technische Erzeuger verschwindet.«[20] Ohne sie beim Namen zu nennen, beschuldigte er

einige »Wilde« der Äußerlichkeiten. Es ist eindeutig, daß er hiermit die verbliebenen Mitglieder der Künstlervereinigung meinte.

In seinem dritten Artikel »Zwei Bilder« stellte er die These auf, daß sich die Welt in einer ähnlichen Umbruchstimmung befinde wie zur Zeit der Christianisierung. Damals sei eine völlig neue Kunst entstanden, die auch erst einmal habe begriffen werden müssen. Er verglich dann eine Märchenillustration mit einem Bild Kandinskys und vertrat die Meinung, daß beide von derselben Innerlichkeit beseelt seien. Das Bild des 19. Jahrhunderts habe allerdings noch jedermann verstanden, das sei in der heutigen Situation nicht mehr so. Die heutige Kunst entstehe derzeit noch ohne eine Verbindung mit dem Publikum, »eher ihrer Zeit zum Trotz«. Diese Bilder »sind eigenwillige, feurige Zeichen einer neuen Zeit, die sich heute an allen Orten mehren. Dieses Buch soll ihr Brennpunkt werden, bis die Morgenröte kommt und mit ihrem natürlichen Licht diesen Werken das gespenstige Aussehen nimmt, in dem sie der heutigen Welt noch erscheint. Was heute gespenstig erscheint, wird morgen natürlich sein.«[21]

»Der Blaue Reiter« wurde in einer so hohen Stückzahl verkauft – 1200 Exemplare waren gedruckt worden –, daß 1914 eine Neuauflage nötig wurde. Finanziell deckte der Verkauf jedoch nicht die Kosten, da das Buch völlig unter Preis angeboten worden war. Das hing auch mit ungenauen Absprachen zwischen den Herausgebern und dem Verleger zusammen. Es war Bernhard Koehler, der dann hilfreich einsprang, da Marc und Kandinsky laut Vertrag mit Piper für die Deckung der Kosten aufkommen mußten.

Ein zweiter Band war bereits bei der Fertigstellung des ersten in Planung. Doch die Arbeiten daran verzögerten sich immer wieder. Um nicht völlig von der Malerei abgehalten zu werden, beschlossen Marc und Kandinsky, abwechselnd für die Herausgabe verantwortlich zu zeichnen. Den zweiten Band hatte Marc übernommen. Dennoch gab es Absprachen zwischen den beiden Künstlern. Im März 1914, als die neuen Vorworte für die zweite Auflage gerade fertig waren, schrieb Kandinsky an Marc, er müsse sich zeitweise völlig aus dem Projekt zurückziehen, da es ihn doch so sehr gefangen nehme, daß er nichts anderes mehr denken könne und seine Malerei darunter leide. Marc war enttäuscht, respektierte aber die Entscheidung des Freundes. Diese Tendenzen kann man auch in den beiden, nach der Entscheidung Kandinskys überarbeiteten Vorworten bemerken: Kandinsky resignativ und Marc hoffnungsvoll die neuen Ideen weiterverfolgend. Aus den Vorworten ging klar hervor, daß so schnell kein neuer Band erscheinen würde. Marc plante allerdings, im Alleingang etwas herauszugeben. Ob er diesen Vorsatz ausgeführt hätte, wenn der Krieg nicht alle Pläne zunichte gemacht hätte, muß dahingestellt bleiben. Es wäre ihm zuzutrauen gewesen, da er in der Zeit zwischen dem Erscheinen des »Blauen Reiters« und dem Ausbruch des Krieges auch andernorts immer wieder in kämpferischer Form für die neue Malerei eingetreten war.

So schrieb er im März 1912 in der Zeitschrift »Pan« den bereits im Zusammenhang mit dem Gemälde *Vögel* von 1914 zitierten Artikel »Die neue Malerei«. Darin versuchte er nachzuweisen, daß diese von ihm als »neue« bezeichnete Malerei nicht die Impressionisten als Vorläufer habe, sondern höchstens, wenn auch nur bedingt, Cézanne. Wichtiger als die Frage nach den Vorläufern erschien ihm diejenige danach, was

Paradies, 1912
zusammen mit August Macke
Öl auf Verputz, 400 × 200 cm
Münster, Westfälisches Landesmuseum für Kunst und Kulturgeschichte

ABBILDUNG SEITE 60/61:
Reh im Klostergarten, 1912
Öl auf Leinwand, 75,7 × 101 cm
München, Städtische Galerie im Lenbachhaus

Skizze von der Brennerstraße, 1913
Bleistift, aquarelliert, 20 × 12,3 cm
München, Galerie Stangl

die neue Kunst wolle: »Wir suchen heute unter dem Schleier des Scheines verborgene Dinge in der Natur... Und zwar suchen und malen wir diese innere, geistige Seite der Natur... weil wir diese Seite sehen, so wie man früher auf einmal violette Schatten und den Aether über allen Dingen ›sah‹. Das Warum können wir für jene so wenig bestimmen wie für uns. Es liegt in der Zeit.«[22] Der letzte Satz hatte für Marc zentrale Bedeutung. Darauf aufbauend führte er weiter aus, daß die neue Malerei »kein Pariser Ereignis« sei, »sondern eine europäische Bewegung«. Er erklärte kategorisch: »Jede Zeit hat ihre Qualität«, und forderte eine sachliche Diskussion »über den künstlerischen Wert oder Unwert der neuen malerischen Ideen«.

Im nächsten Heft erschien eine äußerst polemische Antwort von Max Beckmann, der im Jahr zuvor noch zusammen mit Marc auf den »Protest deutscher Künstler« in der Schrift »Im Kampf um die Kunst« reagiert hatte. Beckmann hatte damals schon von einer anderen Sicht aus argumentiert. Für ihn bedeutete das Wichtigste in der Kunst »die künstlerische Sinnlichkeit, verbunden mit der künstlerischen Gegenständlichkeit und Sachlichkeit der darzustellenden Dinge«. Er führte einen Qualitätsbegriff von Kunst aus, der demjenigen Marcs widersprach. Für Beckmann bedeutete Qualität der Sinn »für den pfirsichfarbenen Schimmer einer Haut, für den Glanz eines Nagels... den Schmelz der Ölfarbe...«. Er polemisierte gegen »eingerahmte Gauguintapeten, Matisse-Stoffe, Picassoschachbrettchen und sibirisch-bajuvarische Marterlnplakate« und endete mit dem Satz: »Die Gesetze der Kunst sind ewig und unveränderlich, wie das moralische Gesetz in uns.«[23]

Marc sah sich gezwungen, darauf zu antworten, obwohl er betonte, daß man auf diese Erwiderung gar nicht antworten könne, daß sie nicht zu einer »eingehenden Diskussion einlade«. Er polemisierte zurück und strich noch einmal seinen Qualitätsbegriff heraus: »mit Qualität bezeichnet man die innere Größe des Werkes, durch die es sich von Werken der Nachahmer und kleinen Geister unterscheidet.«[24]

Hier prallten zwei Richtungen der modernen Malerei aufeinander. Auch Beckmann konnte man ja nicht zu den »ewig Gestrigen« zählen. Vor kurzem sind die beiden Positionen als »Sachlichkeit« und »Innerer Klang« bezeichnet worden. Es ist hier nicht der Ort, zu werten. Tatsache ist, daß sich beide Standpunkte bis heute gehalten haben – und sie bekämpfen sich bis heute mit einer ähnlichen Polemik. Beckmann hat sich dabei als »Wortführer«, als ein Vater der gegenständlichen Kunst behaupten können. Marc hingegen wird von den zeitgenössischen Abstrakten kaum als einer der Ihren angesehen. Seine Popularität als Maler blauer Pferde überdeckt seine kunsttheoretischen Äußerungen und seine Wege zur Abstraktion. Mit dieser Verkürzung wird man dem Weggefährten Kandinskys jedoch nicht gerecht.

Schon 1912, als der »Blaue Reiter« erschien und Beckmann und Marc ihre kontroversen Standpunkte vorführten, zeigten sich bei Marc Tendenzen zur abstrakten Malerei. Die meisten – und die berühmten – seiner Bilder zeigen zwar Tiere, einen Tiger, Kühe oder Pferde. Abstrakte, teils bizarre Formen, die kaum noch Tiefenwirkung vermitteln, sind in dem Gemälde *Reh im Klostergarten* (Abb. S. 60/61) bestimmend. Das Reh – immerhin im Zentrum des Bildes – besteht nur noch aus Rücken und Kopf. Alles andere ist verdeckt, scheint unnötig. Erst durch das wenige, was man von dem Reh sieht, erschließen sich die übrigen For-

Der Mandrill, 1913
Öl auf Leinwand, 91 × 131 cm
München, Staatsgalerie moderner Kunst

men als Bäume, Sonne und Bauten. Sobald etwas Vertrautes in einem Bild erscheint, gruppiert man den Rest darum herum. Es ist allerdings fraglich, ob Marc diese Assoziationen beabsichtigt hat. Zweifellos hat er, wenn auch vielleicht unabsichtlich, durch den Titel den Betrachter aufgefordert, Bauwerke und Pflanzen in seinem Gemälde zu bestimmen. Doch war ihm die Abstraktion andererseits Herausforderung. Das erkennt man auch an seinen »Thesen über die abstrakte Kunst«, die er 1912/13 niederschrieb. Wie bei Kandinsky waren die Vorstellung, der Wille vorher da, bevor sie in die Tat umgesetzt werden konnten.

Für die Entwicklung hin zur Abstraktion und auch schon zu solchen Gemälden wie *Reh im Klostergarten* oder *Im Regen* (Abb. S. 57) war eine Reise im Oktober 1912 von entscheidender Bedeutung. Franz und Maria Marc fuhren zu Mackes nach Bonn, um sich endlich die Sonderbundausstellung in Köln anzuschauen. Im Sommer, kurz vor der Eröffnung, hatte es wegen der Ausjurierung etlicher Bilder der Gruppe, die sich um die Redaktion des »Blauen Reiters« scharte und zu der inzwischen auch Alexej Jawlensky und Marianne von Werefkin gehörten, eine große Auseinandersetzung zwischen Marc und Macke gegeben, da Macke an der Organisation der Ausstellung beteiligt gewesen war. Die Münchner waren deshalb nicht zur Eröffnung gefahren. Doch nun war Marc von einigen Gemälden sehr angetan. Begeistert schrieb er an Kandinsky über

Drei Tiere (Hund, Fuchs und Katze), 1912
Öl auf Leinwand, 80 × 105 cm
Mannheim, Städtische Kunsthalle

Wildschweine (Eber und Sau), 1913
Öl auf Malpappe, 73 × 57,5 cm
Köln, Wallraf-Richartz-Museum

Der blaue Reiter und sein Pferd, 1912
Tinte und Tusche, 15,4 × 11,4 cm
München, Bayerische Staatsgemälde-sammlungen

Werke von Munch, Heckel, Picasso und vor allem Matisse. »Meine Sachen mag ich hier gar nicht, süß und schön, ich bin ganz erschrocken«. Spontan entschlossen sich die Freunde, für eine Woche nach Paris zu fahren, wo sie unter anderem mit Delaunay zusammentrafen. Robert Delaunay (1885–1941) war vom Kubismus geprägt worden. Er löste sich allmählich davon, da er der Farbe immer mehr Bedeutung zumaß, was nicht in das kubistische Konzept paßte. Er begann, sich intensiv mit der Beziehung der Komplementärfarben und mit den Simultankontrasten zu beschäftigen. Simultankontraste entstehen dadurch, daß das Auge zu einer Farbe die Komplementärfarbe verlangt und sie selbst bildet, wenn sie nicht vorhanden ist. Diese Auseinandersetzung endete für Delaunay mit dem Vorrang der Farbe: Die Farbe wurde für ihn zum wichtigsten »Gegenstand« des Bildes, was 1912 zur »reinen Malerei«, der vollkommenen Befreiung vom Gegenstand führte. Der Dichter Guillaume Apollinaire gab dieser Kunstrichtung, die auch Parallelen zu den Gesetzmäßigkeiten der Musik suchte, den Namen »Orphismus«.

Erinnert man sich an die Briefe, in denen sich Marc und Macke im Jahr 1910/11 über Farbprobleme ausgetauscht hatten, ist es verständlich, daß sie von Delaunays Bildern beeindruckt waren. Macke und Delaunay hatten in ihren Theorien etliche Gemeinsamkeiten festgestellt, die sich bei einem Besuch des Franzosen in Bonn in Gesprächen bestätigten. Marc und Delaunay hatten sich jedoch nur wenig zu sagen, und so beschränkte sich auch der Einfluß des Franzosen bei Marc auf gewisse Stilmittel, die er übernahm, und war nicht, wie für Macke, eine Offenbarung.

Wieder zurück in Bonn, hatte Marc die Gelegenheit, im Kölner Gereonsklub die Ausstellung der italienischen »Futuristen« hängen zu helfen. Bislang hatte er nur den Katalog der Ausstellung, die im Frühjahr in Berlin bei Walden in der Galerie »Der Sturm« zu sehen gewesen war, lesen können. Schon die schriftlichen Äußerungen der Italiener hatten ihn sehr beeindruckt. Jetzt begeisterten ihn auch die Bilder. Den Futuristen war es ein Anliegen, Licht, Beleuchtung und Atmosphäre in ihren Gemälden festzuhalten und darüber hinaus Bewegung, Geschwindigkeit, Simultaneität und Durchdringung darzustellen.

Sowohl die Malerei Delaunays als auch die Prinzipien der Futuristen hat Marc in seinen späteren Bildern verarbeitet, ohne dabei »sein« Sujet, das Tier, aus den Augen zu verlieren.

Abgesehen von den vielfältigen Eindrücken dieser Reise konnte Marc in Bonn endlich einen Wunsch verwirklichen, den er schon zu Beginn der Freundschaft mit Macke geäußert hatte. Die Freunde bemalten gemeinsam eine Wand von Mackes Atelier mit dem Paradies-Thema (Abb. S. 58). Allerdings wählten sie nicht den Sündenfall, sondern den paradiesischen Urzustand für das große Wandgemälde. Sie hatten sich künstlerisch noch nicht sehr weit voneinander entfernt, so daß ihnen diese gemeinsame Arbeit noch möglich war.

Weihnachten waren die Marcs wieder in Berlin bei den Eltern von Maria. Dort lernten sie die Dichterin Else Lasker-Schüler kennen, mit der sie bald eine enge Freundschaft verband. Bereits aus München hatte Marc ihr eine von ihm gemalte Postkarte geschickt, um seine Adresse in Berlin mitzuteilen. Neben einem Pferd und einem Mann (Franz Marc) stehen die Worte: »Der blaue Reiter präsentiert Eurer Hoheit sein blaues Pferd...« (Abb. S. 66). Diese Karte war die erste von 28 Kartengrüßen, die der Maler in den folgenden Jahren an die Dichterin schrieb (Abb. S. 68/69). Sie wurden mit langen Briefen beantwortet, die Else Lasker-Schüler auch mit Zeichnungen illustrierte. Diese Briefe bildeten später die Grundlage für ihren Roman »Malik«. Die Korrespondenz ist ein Dokument dafür, wie sich expressionistische Dichtung und Malerei ergänzten. Erst 1987 konnte das gesamte Material veröffentlicht und kommentiert werden.[25] Hier sei darauf hingewiesen, daß der erste farbige Entwurf zum *Turm der Blauen Pferde* 1913 den Neujahrsgruß an Lasker-Schüler darstellte. Mondsichel und Sterne, die die Pferde zieren, waren die persönlichen Insignien der Dichterin. Auf sie und auf ihre Lieblingsfarbe Blau nahm Marc mit diesem Bild Bezug. Er übernahm diese Details in sein großes Gemälde gleichen Titels (Abb. S. 45), das der Berührung mit der Dichterin viel zu verdanken hat.

ABBILDUNGEN SEITE 68:

Der Turm der blauen Pferde, 1912/13
Tusche, Deckfarben, 14,3 × 9,4 cm
München, Bayerische Staatsgemäldesammlungen

Die drei Panther des Königs Jussuf, 1913
Tusche, Aquarell, Deckfarben, 13,8 × 9 cm
München, Bayerische Staatsgemäldesammlungen

Zitronenpferd und Feuerochse des Prinzen Jussuf, 1913
Tusche, Aquarell, Deckfarben, 14 × 9,1 cm
München, Bayerische Staatsgemäldesammlungen

Das Spielpferd des Königs Abigail, 1913
Tusche, Aquarell, Deckfarben, 13,9 × 9 cm
München, Bayerische Staatsgemäldesammlungen

ABBILDUNGEN SEITE 69:

Die Mutterstute der blauen Pferde, 1913
Tusche, Deckfarben, 14 × 9 cm
München, Bayerische Staatsgemäldesammlungen

Das heilige Kälbchen, 1913
Deckfarben, Tusche, 14 × 9 cm
München, Bayerische Staatsgemäldesammlungen

Pferd, 1914
Tusche, Aquarell, Deckfarben, 14,9 × 9,9 cm
München, Bayerische Staatsgemäldesammlungen

Der Traumfelsen, 1913
Tusche, Aquarell, Deckfarben, 14,2 × 8,9 cm
München, Bayerische Staatsgemäldesammlungen

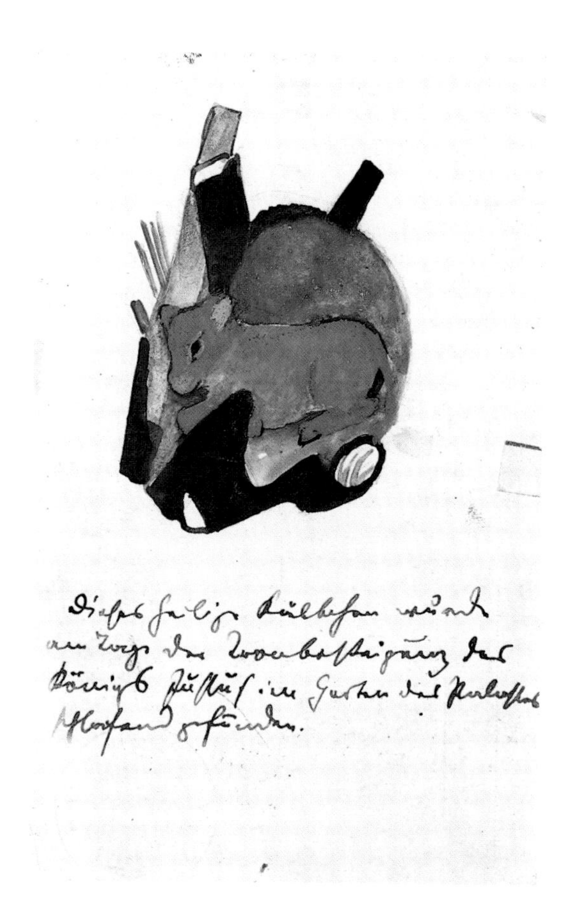

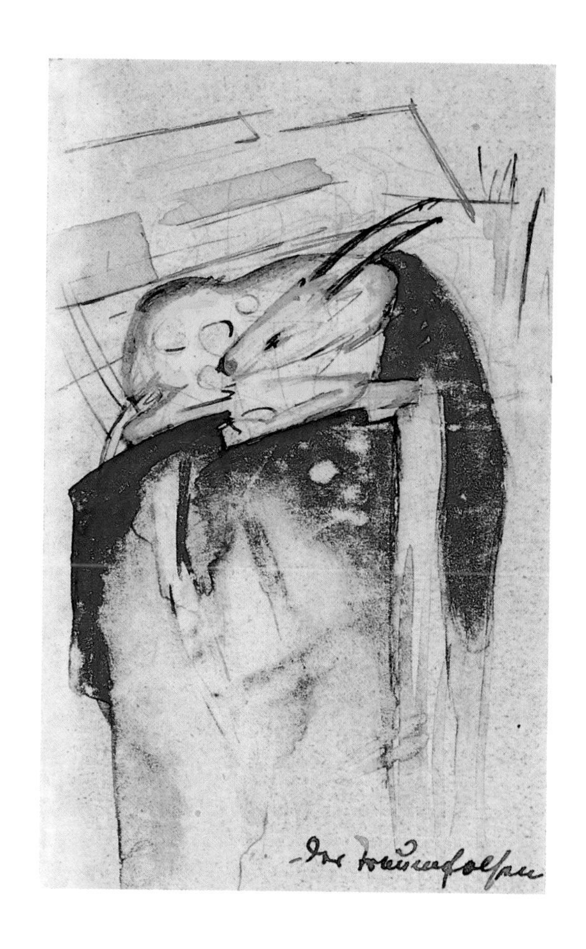

Tierschicksale (Die Bäume zeigten ihre Ringe, die Tiere ihre Adern), 1913
Öl auf Leinwand, 196 × 266 cm
Basel, Kunstmuseum Basel

Füchse, 1913
Öl auf Leinwand, 87 × 65 cm
Düsseldorf, Kunstmuseum

Caliban. Figurine für Shakespeares »Sturm«, 1914
Aquarell und Deckfarben, 46 × 39,7 cm
Basel, Kunstmuseum Basel, Kupferstichkabinett

Kunstpolitik und die malerische Entwicklung bestimmten auch das Jahr 1913. Im Januar veranstaltete die Galerie Thannhauser eine große Ausstellung mit den Werken von Marc, die dann über Jena nach Berlin reiste. Im April bekam Marc an der Stuttgarter Akademie eine Professur angeboten, die er ablehnte.

In diesem Frühjahr faßte er zusammen mit Kandinsky den Plan, eine illustrierte Bibel herauszugeben. Er hatte bei Alfred Kubin, Paul Klee, Erich Heckel und Oskar Kokoschka angefragt, ob sie an diesem Projekt mitwirken wollten, und allgemein Zustimmung erhalten. Jeder Maler konnte selbst entscheiden, welches Kapitel er übernehmen wollte. Kandinsky hatte sich für die Apokalypse entschieden, Marc selber für die Genesis (1. Buch Moses), Kubin arbeitete am Buch Daniel, Kokoschka und Heckel am Buch Hiob, Paul Klee hatte die Psalmen übernommen. Das Ganze sollte als »Blaue-Reiter-Ausgabe« erst in Einzelbänden und zum Schluß noch einmal als Gesamtausgabe erscheinen, deshalb war die Blattgröße für alle Künstler einheitlich festgelegt. Man hatte sich außerdem auf hochformatige Holzschnitte geeinigt, um das lästige Wenden beim Betrachten der Bilder zu vermeiden.

Die Arbeiten der einzelnen Illustratoren gingen nur langsam voran, und Marc sprach erst im Juli 1914 mit Reinhard Piper über diese Idee, als er Kubins Beitrag vorzeigen konnte. Die Verhandlungen mit Piper waren noch nicht sehr weit gediehen, als der Krieg ausbrach. Das Buch Daniel, das als einziges verwirklicht wurde, erschien dann 1918 als Einzelwerk in einem anderen Verlag. Von Franz Marc gab es zu diesem Zeitpunkt nur wenige Holzschnitte zur Schöpfungsgeschichte. Ihren Titeln wie *Geburt der Pferde, Geburt der Wölfe* ist zu entnehmen, daß ihm zahlreiche Illustrationen vorschwebten, deren Motive er sehr frei vom eigentlichen Text ausgewählt hatte. Noch im Krieg dachte er an dieses Projekt. Einige Zeichnungen aus dem Skizzenbuch zeugen davon.

Marcs große kunstpolitische Unternehmung des Jahres 1913 war die Ausrichtung des »Ersten deutschen Herbstsalons« in der Galerie »Der Sturm« in Berlin. Diese Galerie war aus der gleichnamigen literarischen Zeitschrift entstanden, die Herwarth Walden (1878–1941) 1910 gegründet hatte. 1912 entschloß er sich, neben der Zeitschrift, die die modernen Strömungen der Literatur vertrat, eine Galerie zu unterhalten, in der die moderne Kunst ein Forum finden sollte. Mit der ersten Ausstellung der Redaktion des »Blauen Reiters« eröffnete er die Galerie. Als nächstes folgten die Futuristen. Im Herbst präsentierte er die Zurückgewiesenen der Sonderbundausstellung in Köln. Den von Marc gewünschten Titel »Die Refüsierten« wandelte er in »Deutsche Expressionisten« um. Zum erstenmal wurden die Maler um den »Blauen Reiter« so bezeichnet. Später ging dieser Titel an die »Brücke«-Maler über.

Die Art und Weise, wie bei der Sonderbundausstellung in Köln die Juroren mit den Werken der Maler umgegangen waren – August Macke hat dies später in einem Brief an Marc bestätigt –, und die Erfahrungen mit der daraus hervorgehenden Ausstellung der »Deutschen Expressionisten« in Berlin waren ein Grund für die Ausrichtung des »Ersten Deutschen Herbstsalons«. Weitere Gründe waren Marcs und Mackes Unzufriedenheit mit dem wenig lebendigen »Salon d'Automne« 1912 in Paris sowie der Wunsch, dem einflußreichen Berliner Kunsthändler Paul Cassirer bei einem ähnlichen Projekt zuvorzukommen. Neben dem Veranstalter Herwarth Walden waren vor allem Macke und die »Blauen Rei-

ter« Marc und Kandinsky an der Organisation und der Auswahl der Künstler beteiligt. Klee, Delaunay und andere wurden beratend hinzugezogen. Sie sollten sich vor allem um geeignete Künstler ihrer Nationalität kümmern, in diesem Fall um Schweizer und Franzosen.

90 Künstler und Künstlerinnen aus Frankreich, Deutschland, Rußland, Holland, Italien, Österreich, der Schweiz und den USA bestritten mit 366 Werken den »Ersten Deutschen Herbstsalon«. Besonders viel Platz wurde Robert Delaunay und seiner Frau Sonja Delaunay, den Mitorganisatoren Marc, Macke und Kandinsky, den Künstlern um den »Blauen Reiter«, wie Heinrich Campendonk, Gabriele Münter, Paul Klee und Alfred Kubin, und den italienischen Futuristen eingeräumt.

Rückblickend ist der Herbstsalon eine der bedeutendsten Übersichtsausstellungen der modernen Kunst vor dem ersten Weltkrieg gewesen. Walden hatte seinen Anspruch, die Avantgarde in Deutschland zu vertreten, einlösen können. Zu ihrer Zeit begleiteten jedoch Hetzkampagnen der Presse die Ausstellung. Auch finanziell wurde der Herbstsalon ein Desaster. Bernhard Koehler, der eine Ausfallgarantie von 4000 Mark übernommen hatte, mußte diese am Ende der Ausstellung auf 20000 Mark erhöhen. Marc zeigte im Herbstsalon sieben neue Gemälde, darunter den *Turm der Blauen Pferde*, die *Tierschicksale*, *Tirol*, *Die ersten Tiere* und *Die Wölfe (Balkankrieg)*. Im Mai hatte er in einem Brief an Macke diese Titel und noch einige andere wie *Das arme Land Tirol* und *Die Weltenkuh* angekündigt und geschrieben, daß es lauter ganz

Das arme Land Tirol, 1913
Öl auf Leinwand, 131,5 × 200 cm
New York, Solomon R. Guggenheim Museum

Fabeltier II (Pferd), 1913
Tempera auf Karton, 26,5 × 30,5 cm
Privatbesitz

Vögel, 1914
Öl auf Leinwand, 109,5 × 99,5 cm
München, Städtische Galerie im Lenbachhaus

verschiedene Bilder seien, die er gerade fertiggestellt habe. Diese Feststellung entsprach durchaus den Tatsachen.

In dem Gemälde *Das arme Land Tirol* (Abb. S. 73), dem er kurze Zeit später noch das Bild *Tirol* (Abb. S. 77) folgen ließ, gab er seine Eindrücke von einer Reise mit Maria Marc nach Südtirol im März 1913 wieder. Zwischen hoch aufragende Berge verteilte Marc Häuser, Grabkreuze und einige wenige Tiere. Abgesehen von der Farbigkeit arbeitete er hier mit erstaunlich realistischen Versatzstücken, die gleichsam wie eine Zurücknahme der Auffassungen wirken, die er in Gemälden wie *Reh im Klostergarten* (Abb. S. 60/61) ein Jahr zuvor noch zum Ausdruck gebracht hatte. Dagegen wirkt der *Turm der Blauen Pferde* (Abb. S. 45) sehr viel stärker abstrahiert. Die Körper der Pferde setzen sich aus einzelnen geometrischen Teilen zusammen, die »Landschaft« besteht nur noch aus abstrakten Formationen. Diese Abstraktion ist in dem Bild *Tierschicksale* (Abb. S. 71) auf die Spitze getrieben. Mühsam schälen sich aus den spitzen, bedrohlich erscheinenden Formen Pferde, Schweine, Wölfe und ein Reh heraus. Die relative Ruhe, die in den beiden zuerst genannten Gemälden vorherrscht, ist hier einer enormen Dynamik gewichen, die Form und Tier ergriffen hat. Es scheint, als wolle sich die Natur in einer selbstzerstörerischen Kraft aufbäumen.

Marc hatte dem Gemälde erst den Titel *Die Bäume zeigten ihre Ringe, die Tiere ihre Adern* gegeben, auf der Rückseite vermerkte er »Und alles Sein ist flammend Leid«. Es war Klee, der ihm auf die Frage nach einem geeigneten Namen »Tierschicksale« vorschlug. An das Gemälde knüpfen sich wie an den *Turm der Blauen Pferde* viele Spekulationen. Marc selber schrieb 1915 an Maria, als er von Koehler eine Kunstpostkarte mit den *Tierschicksalen* erhalten hatte: »Es ist wie eine Vorahnung dieses Krieges, schauerlich und ergreifend; ich kann mir kaum vorstellen, daß ich das gemalt habe! In der verschwommenen Photographie wirkt es jedenfalls unfaßbar wahr, daß mir ganz unheimlich wurde.«[26]

Was lag näher, als Marc prophetische Gaben zuzusprechen. Einer Interpretation aus dem Jahr 1956 zufolge sind in dem Bild vier menschliche Verhaltensweisen angesichts einer hereinbrechenden Katastrophe zu erkennen. Wahrscheinlicher ist die Annahme, daß Marc hier eine ganz persönliche Apokalypse stattfinden ließ. Er mußte seine Tiere preisgeben, um sich von ihnen zu befreien, um zu dem gegenstandslosen Bild vordringen zu können. Die Gleichzeitigkeit von Gemälden wie *Turm der Blauen Pferde* und *Tierschicksale* mutet dabei seltsam an. Sie ist Ausdruck einer Ambivalenz, die 1914 noch stärker hervortrat und sowohl in den gegenständlichen als auch in den sogenannten abstrakten Werken Marcs zu finden ist.

Marc hat an der Wichtigkeit des Gegenstandes in einem Bild immer festgehalten, obwohl er die abstrakte Malerei von Kandinsky und Delaunay verteidigte und 1913 selber sein erstes abstraktes Gemälde schuf. Er faßte seine Auffassung in einem Brief an Walden während der Vorbereitungen für den Herbstsalon in Worte. Walden hatte offensichtlich das »Gegenständliche« in der Kunst für belanglos erklärt. Vehement setzte Marc dagegen: »Ob man eine thronende Maria oder Spargeln malt, ist nicht ausschlaggebend für die Qualität des Bildes und seinen Wert, kann aber doch einen *verdammten* Unterschied bedeuten...«[27] Dieser »Unterschied« wird vor allem in den letzten Werken spürbar.

Tirol, 1914
Öl auf Leinwand, 135,7 × 144,5 cm
München, Staatsgalerie moderner Kunst

Gazellen, 1913
Tempera, 55,5 × 71 cm
Krefeld, Privatbesitz

Rehe im Walde II, 1914
Öl auf Leinwand, 110 × 100,5 cm
Karlsruhe, Staatliche Kunsthalle

Letzte Werke – Krieg und Tod

Nach dem Herbstsalon, der Marc durch die Werke Delaunays und der Futuristen, durch die Diskussionen mit Herwarth Walden und anderen Künstlern neue Impulse gegeben hatte, entstanden zunächst Gemälde wie *Stallungen* (Abb. S. 86) und *Bild mit Rindern* (Abb. S. 83). Beide Gemälde sind geometrisch aufgebaut. Rechtwinklige Formen dominieren. Weder die Pferde in *Stallungen*, noch die Rinder im zweiten Bild wirken organisch, sondern »gebaut«, konstruiert. Futurismus und Orphismus gehen eine Verbindung ein.

Sie führten zu den *Kleinen Kompositionen*, deren erste (Abb. S. 84) Marc im Dezember 1913 schuf. Kurz vorher hatte er eine Postkarte für Lily Klee gemalt, die als sein erster abstrakter Versuch gilt. Abgesehen von der *Kleinen Komposition II* (Abb. S. 85), welche auch den Untertitel *Haus mit Bäumen* trägt, entstanden noch zwei weitere gegenstandslose Gemälde, die Marc als *Kleine Kompositionen* (Abb. S. 87) bezeichnete und die dem Orphismus Delaunays Rechnung tragen. Doch wie schon in der Zeit vor dem Herbstsalon ist weiterhin kein einheitlicher Stil bei Marc festzustellen. Gleichzeitig mit den eben erwähnten Bildern malte er die *Vögel* (Abb. S. 75), die stilistisch weder den letzten Gemälden von 1913 noch den *Kompositionen* zugeordnet werden können.

Diese Gemälde sind offensichtlich alle noch in Sindelsdorf entstanden, doch bestand bereits die Aussicht auf einen eigenen Besitz, die sich für die Marcs relativ überraschend ergeben hatte. Durch einen Brief von Maria Marc an Lisbeth Macke sind die Details bekannt.

Im Januar 1914 hatte das Ehepaar Marc begonnen, sich in der Nähe von Kochel, wo Franz Marc früher häufig die Sommer zugebracht hatte, nach einem Haus umzusehen. Die Sindelsdorfer Bauern konnten sich auf die Dauer wohl nicht so recht mit den anderen Lebensgewohnheiten der Künstler abfinden, oder auch die Maler sich nicht dem bäuerlichen Leben anpassen. Auch Niestlés zogen an den Starnberger See. Von der einstigen kleinen Künstlerkolonie blieben nur Campendonk und seine Frau in Sindelsdorf zurück.

Der Zufall wollte es, daß Franz und Maria Marc in Ried bei Benediktbeuren ein großes Haus fanden, das zum Verkauf angeboten wurde und dessen Besitzerin in einen Münchner Vorort ziehen wollte. Franz Marc hatte inzwischen das Elternhaus in Pasing geerbt und bot dieses zum Tausch an. Mit Unterstützung der Schwiegermutter erstanden sie auch noch etwas Land, da der Garten nicht sehr groß war und die Rehe, die er sich inzwischen gekauft hatte, ein großes Gehege brauchten. Auch ein Atelier sollte angebaut werden, doch dazu kam es nicht mehr.

Skizzenbuch aus dem Felde, 1915
Bleistift, je 16 × 9,8 cm resp. 9,8 × 16 cm
München, Staatliche Graphische Sammlung
Blatt 9: *sehr farbig*

Zerbrochene Formen, 1914
Öl auf Leinwand, 112 × 84,5 cm
New York, Solomon R. Guggenheim Museum

Kurz nach dem Brief von Maria Marc an Lisbeth Macke, den sie im Februar 1914 schrieb, zeichnete Franz Marc für August Macke das neue Anwesen und gestand: »Malen tu ich momentan nicht viel, – ich hab den Kopf zu voll mit diesem Lokalwechsel. Dort wird die Geburtsstätte des 2. Blauen Reiter-Bandes, – lach nur; er wird doch! Ich brüte wie eine Henne darüber; ...«[28]

Seine Zuversicht wurde durch die bereits erwähnte Absage Kandinskys gebremst. Kurze Zeit später, noch vor dem Umzug nach Ried Ende April, bekam Marc vom damaligen Dramaturgen der Münchner Kammerspiele Hugo Ball (1886–1927) das Angebot, Shakespeares »Sturm« zu inszenieren. Außerdem wollte Ball ein Buch über »Das neue Theater« herausbringen. Sowohl die Autoren als auch der Verlag waren dieselben wie beim »Blauen Reiter«. Marc war von der Idee begeistert und versuchte, für seine Inszenierung Musik von Schönberg selber oder von dessen Schülern Anton von Webern oder Alban Berg zu erhalten. Mit diesem Anliegen schrieb er am 9. April an Kandinsky. Doch bereits am 18. desselben Monats resignierte er, wie einem Brief an Hugo Ball zu entnehmen ist: »... den Zeitungsnotizen nach zu schließen scheint unser Theaterprojekt in den üblichen Münchner goldenen Mittelweg auszulaufen; die Zeitungen reden um den Begriff »Münchner Künstler« herum... Es müßte doch unbedingt ausgesprochen werden, ... daß wir die Scene selbst... neu organisieren und nach unserem künstlerischen Vorstellungsleben gestalten wollen.« Er sah den Konflikt mit den Schauspielern voraus, der einem blühe, wenn man nicht nur schöne Dekorationen schaffe, wie er sie anderen Künstlern, etwa Jawlensky, zutraute. Er wollte lieber noch warten, bis wirklich etwas Neues mit »eigenem Ensemble

Liegender Stier, 1913
Tempera, 40 × 46 cm
Essen, Museum Folkwang

Bild mit Rindern, 1913
Öl auf Leinwand, 92 × 130,8 cm
München, Staatsgalerie moderner Kunst

und vollkommener Bewegungsfreiheit« möglich sei. »Dann müßte man Kokoschka, Kandinsky, Klee, Macke berufen und als Musiker... den Schönberg-Kreis... gewinnen, alles zeitig vorbereitet, damit nichts überstürzt wird...«[29]

An dem Buch, das im Oktober erscheinen sollte, arbeitete er jedoch weiterhin mit. Der fragmentarische Aufsatz »Das abstrakte Theater« und zwei Aquarelle (Abb. S. 72) zeugen davon.

In Ried entstanden die vier Gemälde *Heitere Formen* (im zweiten Weltkrieg zerstört), *Spielende Formen* (Abb. S. 88/89), *Kämpfende Formen* (Abb. S. 91) und *Zerbrochene Formen* (Abb. S. 80). Schon an den Titeln erkennt man die Zwiespältigkeit der Gefühle, die den Maler damals bewegten. In den ersten beiden Bildern werden die Idylle, die Harmonie, die positiven Seiten des Lebens beschworen, in den zwei anderen Kampf und Untergang. Das Bild *Kämpfende Formen* wurde unlängst als »Kampf spiritueller Kräfte gegen die materielle Welt«[30] gedeutet. Im zeitgeschichtlichen Zusammenhang gesehen scheint jedoch auch Marcs persönliches Leben in Ried – die Idylle – zu der damals herrschenden Weltuntergangsstimmung, die dann im ersten Weltkrieg mündete, in Widerspruch zu stehen und mit ihr zu konkurrieren. Abgesehen von diesen allgemeinen, inhaltlichen Interpretationen ist interessant, daß Marc die Formen gleichsam figürlich auffaßte und als Subjekte begriff. Damit

Kleine Komposition I, 1913
Öl auf Leinwand, 46,5 × 41,5 cm
Schweiz, Privatbesitz

Kleine Komposition II (Haus mit Bäumen), 1914
Öl auf Leinwand, 59,5 × 46 cm
Hannover, Sprengel Museum

nahm er die Abstraktion wieder zurück. In dem Gemälde *Kämpfende Formen* kann – Levine zufolge – das rote Feld links im Bild als Adler gedeutet werden, der sich auf ein blau-schwarzes, nicht näher zu definierendes Wesen stürzt.

Die Ambivalenz der Gefühle tritt in den beiden »gegenständlichen« Bildern, die als letzte in Ried entstanden bzw. umgearbeitet wurden, deutlicher zu Tage: *Tirol* (Abb. S. 77) hatte Marc gleichzeitig mit dem Gemälde *Das arme Land Tirol* (Abb. S. 73) gemalt, kurz nach seiner Reise in dieses Land. Es war auch kurz im Herbstsalon ausgestellt, Marc hatte es dann zurückgenommen. Er überarbeitete es und fügte die Madonna auf der Mondsichel im Zentrum des Bildes hinzu. Sie steht inmitten der hoch aufragenden Berge, die bedrohlich spitz nach oben zulaufen. Von der Muttergottes gehen Strahlen aus, die diese Berge rechtwinklig durchschneiden. Ganz unten ducken sich wenige Häuser in der Ebene, bedroht von den Bergen und einem diagonal ins Bild hineinragenden Baum, der in seiner Form an eine Sense erinnert. Die aufgehende Sonne rechts oben hinter den Bergen ist verdunkelt, ihr stehen links oben gleich mehrere Mondsicheln gegenüber.

Sense und Madonna erinnern an Motive in christlichen Darstellungen der Apokalypse, die der »Offenbarung des Johannes« entnommen sind. Die Untergangsstimmung wird vor allem durch die 1914 hinzugefügte Madonna direkter ins Bild umgesetzt als in den vorher entstandenen abstrakten Gemälden. Hinzu kommt eine religiöse Auffassung, die derart thematisch im Œuvre Marcs, außer in den Bibelillustrationen, nicht zu erkennen war und die an seine christlich geprägte Jugend erinnert. Die »andere Seite«, die Idylle, griff Marc in seinem letzten Gemälde *Rehe im Walde II* (Abb. S. 79) noch einmal auf. Das Glück wird hier in einem Familienbild überhöht, in dem auch die Charaktere der Farben noch einmal, wenn auch abgewandelt, hervortreten.

Auf einer – natürlich stark abstrahierten, aus den Primär- und Sekundärfarben zusammengesetzten – Waldlichtung lagern drei Rehe.

Stallungen, 1913
Öl auf Leinwand, 73,5 × 157,5 cm
New York, Solomon R. Guggenheim Museum

Kleine Komposition III, 1913/14
Öl auf Leinwand, 46,5 × 58 cm
Hagen, Karl Ernst Osthaus Museum

Der große Bock, der durch sein leuchtendes Blau dominiert (Blau als männliches Prinzip), liegt im Zentrum des Bildraumes. Den Kopf hat er witternd zurückgewendet, was seine Größe noch unterstreicht. Kopf und Leib bilden so ein großes, das Bild beherrschendes Dreieck. Links unten lagert die rote Ricke. Die Farbe Rot bedeutete für Marc die Materie. 1910 hatte er sie als brutal und schwer charakterisiert. Hier vergegenwärtigte sie nun die Erde, das Weibliche, welches Leben hervorbringt. Die Ricke liegt vom Betrachter abgewandt und blickt zu ihrem Kitz, welches den Blick erwidert. Dieses Kind kann aufgrund der Farbe Gelb als sanft, heiter und sinnlich angesehen werden. Die Familie bildet einen Kreis, aus dem alle anderen - auch die Betrachter - ausgeschlossen sind. Die Funktionen sind durch Farben und Gehabe klar definiert: Der beschützende Mann, die umsorgende Frau und das auf die Mutter ausgerichtete Kind.

Marc wiederholt hier in seinem Tierbild die Formeln, die seit Jahrhunderten das Familienbild prägten und vor allem durch Gemälde von der »Heiligen Familie« immer wieder reproduziert wurden. Es umfaßt offensichtlich auch das, was Marc unter »Glück« verstand.

Spielende Formen, 1914
Öl auf Leinwand, 56,5 × 170 cm
München, Privatbesitz

Am 1. August 1914 erklärte das Deutsche Reich Rußland den Krieg und gab den Befehl zur allgemeinen Mobilmachung. Franz Marc und August Macke gehörten zu den Unzähligen, die sich sofort freiwillig meldeten.

Diese Kriegsbegeisterung, die vor allem die Intellektuellen ergriff, ist heute schwer nachvollziehbar, zumal es auch andere Stimmen gab. Kandinsky hatte schon 1912 in einem Brief an Marc Angst vor dem Krieg geäußert. Ihm war diese Vorstellung schrecklich. Er sprach von den schmutzigen Folgen, deren stinkende Schleppe über den Erdball gezogen würde und beschwor Berge von Leichen. 1914 hatte er seine Auffassung nicht im geringsten geändert.

Lisbeth Macke hingegen schrieb gutgläubig an Maria Marc: »Wie schnell kam alles, und doch eigentlich atmet man jetzt erleichtert auf, gegen die Tage, wo es wie ein Gewitter in der Luft schwebte. Und wenn man sieht, wie gern alle geben, das ist herrlich.«[31]

Marcs Beweggründe dafür, sich freiwillig zum Kriegsdienst zu melden und den Krieg gutzuheißen, sind vor allem idealistischer Natur. Aus dem Feld schickte er Aufsätze an Maria Marc, mit der Bitte, sich um ihre Veröffentlichung zu kümmern. In diesen Texten wird, ebenso wie in seinen Briefen an seine Frau deutlich, daß er ein krankes Europa sah, das nur durch den Krieg geläutert werden könne. Er sprach von einem völkergemeinschaftlichen Blutopfer, durch das die Welt rein werde. Die

Auffassung, daß wirtschaftliche Interessen zu diesem Krieg geführt haben könnten, lehnte er strikt ab. Er begriff den Krieg als einen Bürgerkrieg, einen »Krieg gegen den inneren, unsichtbaren Feind des europäischen Geistes«. Andererseits hatte er auch die Vision, daß Deutschland gestärkt aus diesem Krieg hervorgehen werde, und stellte sich ein Europa unter deutscher Vormachtstellung vor: »Das Deutschtum wird nach diesem Krieg über alle Grenzen schwillen. Wenn wir gesund und stark bleiben und die Frucht unseres Sieges nicht verlieren wollen, brauchen wir... einen Lebensstrom, der alles durchdringt, ohne Angst... vor dem Fremden,... das uns unsre Machtstellung in Europa bringen wird.«[32]

Auch der Tod Mackes im Oktober 1914 konnte seine Einstellung nicht ändern, obwohl er persönlich unter dem Verlust des Freundes litt. Ein spontan geschriebener Nachruf, der erst nach dem Krieg veröffentlicht und viel zitiert wurde, drückt nicht nur die Trauer um den toten Freund aus, sondern hält auch die Notwendigkeit des Opfers fest: »Das Blutopfer, das die erregte Natur den Völkern in großen Kriegen abfordert, bringen diese in tragischer, reueloser Begeisterung. Die Gesamtheit reicht sich in Treue die Hände und trägt stolz unter Siegesklängen den Verlust.«[33]

In seinen bereits 1920 veröffentlichten und berühmt gewordenen »Briefen aus dem Feld« klingen diese Gedanken immer wieder an,

Schöpfungsgeschichte I, 1914
Holzschnitt, 23,8 × 20 cm
München, Staatliche Graphische Sammlung

besonders als Erwiderungen auf die leider nicht veröffentlichten Briefe von Maria Marc, die dem Krieg offensichtlich nicht diese guten Seiten abgewinnen konnte. Den Briefen kann man aber auch entnehmen, womit sich der Maler in dieser Zeit geistig auseinandersetzte. In seinem Feldgepäck hatte er nur das Evangelium des Markus mitgenommen; bald verlangte er aber nach anderen Büchern und erörterte deren Inhalt dann brieflich mit Maria. Seine Gedanken kreisten immer wieder um die Kunst. Die französischen Dörfer, durch die er ritt, verglich er mit impressionistischen Gemälden und Bildern van Goghs. Von einem zusammenfassenden Rückblick über seine künstlerischen Bestrebungen, in dem er die Gründe für seine Zuwendung zum Tier und seine spätere Hinwendung zur abstrakten Kunst benannte, war bereits im Kapitel »Das Bild des Tieres« die Rede.

Seine Gedanken wanderten aber auch immer wieder nach Ried, zu dem Haus und den Tieren. Sogar über ihre richtige Fütterung gab er Ratschläge, und er war untröstlich, als eines der Rehe starb.

Mit der Zeit änderte sich allerdings seine Einstellung zum Krieg, ein Sinneswandel, der ja auch bei vielen anderen zu beobachten war. Max Beckmann mag hier stellvertretend genannt werden. Bereits im Oktober 1915 schrieb Marc an Lisbeth Macke einen Brief, in dem er den Krieg als den »gemeinsten Menschenfang, dem wir uns ergeben haben« bezeich-

»Beständige Meditation über die Form, beständigen Willen zur Form, den man immer wieder korrigirt, verwirft, neu ansetzt, mit allen Hebeln der Welt und Erfahrung, – ohne das geht's nicht. Bloß leben, das Leben fühlen bis zum Kern u. auf die Form warten wie die Blumen auf den Frühling, das war u. ist nie produktive Kunst; das *Werk* freilich muß den dornenvollen Weg ganz vergessen machen. Der Beschauer soll u. kann nur das reine Werk sehen, unsre Nöte gehen ihn nichts an, auch unsre ›Mittel‹ nicht.«

Franz Marc, 29. 3. 1915

Kämpfende Formen, 1914
Öl auf Leinwand, 91 × 131 cm
München, Staatsgalerie moderner Kunst

nete. Maria gegenüber wurde er Neujahr 1916 noch deutlicher: »Die Welt ist um das blutigste Jahr ihres vieltausendjährigen Bestehens reicher. Es ist fürchterlich dran zu denken; und das alles um *nichts*, um eines Mißverständnisses willen, aus Mangel, sich dem Nächsten menschlich *verständlich* machen zu können! Und das in Europa!! Man muß wirklich alles umlernen, neudenken, um mit dieser ungeheuerlichen *Psychologie der That* fertig zu werden und sie nicht nur zu hassen, zu beschimpfen und zu verhöhnen oder zu beweinen, sondern ursächlich zu begreifen und – *Gegengedanken* zu bilden.«[34]

Anfang 1916 gab es eine Hoffnung, daß Marc bald heimkehren würde. In einem Brief vom 4. März schrieb er an Maria: »...ja, dieses Jahr werde ich auch zurückkommen in mein unversehrtes liebes Heim, zu Dir und zu meiner Arbeit. Zwischen den grenzenlosen schaudervollen Bildern der Zerstörung, zwischen denen ich jetzt lebe, hat dieser Heimkehrgedanke einen Glorienschein, der gar nicht lieblich genug zu beschreiben ist.«[35]

An diesem Nachmittag wurde Franz Marc bei einem Erkundungsgang von einem Granatdoppelschuß getroffen. Am nächsten Morgen setzte man den Leichnam im Garten des Schlosses von Gussainville bei. 1917 ließ ihn Maria nach Kochel überführen, wo er seither begraben liegt.

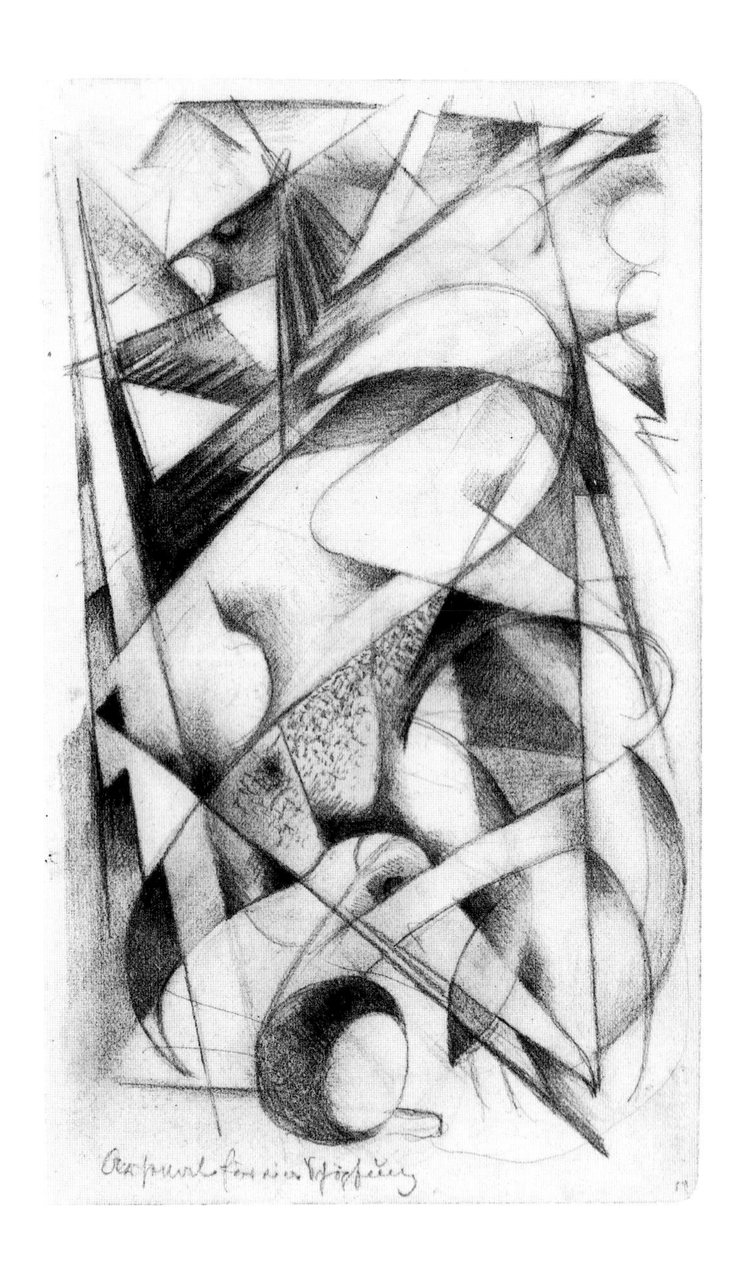

Skizzenbuch aus dem Felde, 1915
Bleistift, je 16 × 9,8 cm resp. 9,8 × 16 cm
München, Staatliche Graphische Sammlung
Blatt 12: *Arsenal für eine Schöpfung*
Blatt 20: *ohne Titel*
Blatt 21: *Zaubriger Moment*
Blatt 30: *Fragment*

Else Lasker-Schüler dichtete auf den Freund einen Nachruf:

> »Als der blaue Reiter war gefallen…
> Griffen unsere Hände sich wie Ringe; –
> Küßten uns wie Brüder auf den Mund.
> Harfen wurden unsere Augen,
> Als sie weinten: Himmlisches Konzert.
> Nun sind unsere Herzen Waisenengel.
> Seine tiefgekränkte Gottheit
> Ist erloschen in dem Bilde: Tierschicksale.«[36]

Im Herbst 1916 veranstaltete die »Münchener Neue Secession« eine Gedächtnisausstellung, was insofern erstaunlich ist, als Marc sich der Mitgliedschaft immer entzogen hatte. Es war der umfassendste Überblick über seine Werke, der jemals gezeigt wurde. Auch Walden stellte noch 1916 in der Galerie »Der Sturm« Werke des Freundes aus dem Besitz von Maria Marc und Bernhard Koehler aus.

Nach dem Krieg begannen die Museen auf Marc aufmerksam zu werden. Der bedeutendste Ankauf war wohl derjenige des *Turm der*

Blauen Pferde 1919 durch die Berliner Nationalgalerie. In den zwanziger Jahren setzte eine Anerkennung des Expressionismus ein, durch die auch das Werk von Franz Marc eine Aufwertung erfuhr. Nach 1933 galt der Künstler – trotz des »Heldentodes« – als entartet. Viele seiner Werke aus öffentlichem Besitz wurden beschlagnahmt und ins Ausland verkauft. Bedeutende Teile der Sammlung Koehler verbrannten 1945 bei der Bombardierung Berlins. Der *Turm der Blauen Pferde*, der 1937 noch in der Ausstellung »Entartete Kunst« zu sehen war, ist seit 1945 verschollen.

Nach dem zweiten Weltkrieg begann der Siegeszug seiner Malerei, der seinen Bildern bald eine inflationäre Verbreitung beschied. Sogar die Deutsche Bundespost brachte eine Briefmarke mit einem »Blauen Pferd« heraus. Über der Flut von Reproduktionen seiner Gemälde vornehmlich aus den Jahren 1911/12 geriet die Bedeutung des Malers als eines Vorreiters der abstrakten Kunst in Vergessenheit. Erst die große Monographie von Klaus Lankheit von 1976 und die Gedächtnisausstellung 1980 im Münchner Lenbachhaus brachten die eigentliche Bedeutung von Franz Marc, die weit über das Malen blauer Pferde hinausreicht, wieder ans Tageslicht.

Leben und Werk im Überblick

1880 Am 8. Februar wird Franz Moriz Wilhelm Marc in München geboren. Er ist das zweite und letzte Kind des Malers Wilhelm Marc und seiner Frau Sophie, geborene Maurice, deren Eltern aus Frankreich stammten. Der Bruder Paul ist drei Jahre älter.

1894 Gemäß der protestantischen Konfession der Mutter wird Marc konfirmiert. Der Konfirmandenunterricht bei dem Pfarrer Otto Schlier hinterläßt bei ihm einen nachhaltigen Eindruck.

1897 Marc ist sich über seinen späteren Werdegang unschlüssig. Theologie, Philologie und Malerei interessieren ihn gleichermaßen. Unter dem Einfluß von Schlier entschließt er sich, Pfarrer zu werden.

1898 Er fühlt sich dem Anspruch, den er selber an einen Pfarrer stellt, nicht gewachsen und erklärt deshalb im Dezember, er ziehe es vor, Philologie zu studieren und Gymnasialprofessor zu werden.

1899 Marc schreibt sich als Student an der Philosophischen Fakultät der Ludwig-Maximilian-Universität in München ein. Bevor er mit dem Studium beginnt, leistet er seinen Militärdienst ab.

1900 Während seiner Militärzeit wird ihm klar, daß er Maler werden will. Im Herbst schreibt er sich an der Münchener Akademie ein. Seine Lehrer sind Gabriel Hackl und Wilhelm von Dietz, beide der Tradition der Münchener Malerei des 19. Jahrhunderts verpflichtet.

1901 Franz und Paul Marc, der in Florenz Byzantinistik studiert, reisen zusammen nach Venedig, Padua und Verona.

1902 Den Sommer verbringt Marc auf der Staffelalm, oberhalb von Kochel.

1903 Marc reist mit dem Studienkollegen Friedrich Lauer nach Frankreich. Er lernt Werke von französischen Impressionisten und von Courbet und Delacroix, sowie gotische Architektur und außereuropäische Kunst kennen. Die Erwerbung einiger japanischer Holzschnitte sind der Beginn einer Ostasiatica-Sammlung. Marc entschließt sich, nicht mehr an die Akademie zurückzukehren.

1904 Marc findet in Schwabing in der Kaulbachstraße ein Atelier. Er befreundet sich mit der verheirateten Malerin und Kopistin Annette von Eckardt. Ihre Ehe belastet die Beziehung. Auch mit seiner Malerei ist der Künstler unzufrieden.

1905 Marc zieht sich immer mehr in sich selbst zurück. Trotzdem fallen in dieses Jahr entscheidende Begegnungen. Er trifft den französischen Tiermaler Jean Bloé Niestlé. Außerdem lernt er die Malerinnen Marie Schnür und Maria Franck kennen. Am Ende des Jahres findet die Beziehung zu Annette von Eckardt ein Ende.

1906 Marc begleitet seinen Bruder auf dessen Einladung hin im April für drei Wochen auf den Berg Athos. Von Mai bis Oktober hält er sich in Kochel auf und malt. Die Malerinnen Schnür und Franck besuchen ihn dort. Das Gemälde *Zwei Frauen am Berg* legt davon Zeugnis ab. An Weihnachten malt er den vom Tod gezeichneten Vater, der kurze Zeit später stirbt.

Wilhelm Marc, Franz Marc am Basteltisch, um 1895
Öl auf Leinwand (Ausschnitt)
Kochel am See, Franz Marc-Museum

1907 Marc heiratet im März Marie Schnür, um ihr zu ermöglichen, ihr uneheliches Kind zu sich zu nehmen. Noch am Abend der Hochzeit fährt er allein nach Paris, wo ihn van Gogh und Gauguin begeistern.
Nach der Rückkehr nach München stellt sich die Heirat mit Marie Schnür als problematisch heraus. Franz Marc zieht sich nach Kloster Indersdorf zurück, um zu malen. Ende Juni bezieht er ein neues Atelier in der Schellingstraße. Ende des Jahres entschließt er sich, seine finanzielle Situation mit Kursen über Tieranatomie-Zeichnen aufzubessern. Der Erfolg ist allerdings gering.

1908 Die Ehe zwischen Marie Schnür und Franz Marc wird geschieden. Allerdings klagt Schnür entgegen ihrer Abmachung auf Ehebruch und verhindert so eine Heirat mit Maria Franck. Im Sommer ist er zusammen mit Maria in Lenggries; sie malen im Freien. Hier entsteht neben *Lärchenbäumchen* vor allem das erste Lenggrieser Pferdebild.

1909 Marc ist von einer Marées-Ausstellung sehr beeindruckt. Die beiden wichtigen Münchener Kunsthändler Thannhauser und Brakl kaufen Arbeiten von ihm. Leben kann er von seiner Kunst jedoch immer noch nicht. Den Sommer verlebt er mit Maria in Sindelsdorf.
Im Dezember wird in München eine Ausstellung von van Gogh ausgerichtet, die Marc hängen hilft. Der Einfluß van Goghs auf seine Malerei wird immer stärker. Ein Beispiel dafür ist *Katzen auf rotem Tuch.*

1910 Anfang Januar lernt Marc August Macke kennen. Es ist für ihn der erste Kontakt mit einem gleichgesinnten Künstler.

Ihre Freundschaft ist schnell besiegelt. Durch Macke wird der Berliner Fabrikant Bernhard Koehler auf Marc aufmerksam. Er kommt im Februar nach München, um sich die erste Ausstellung Marcs in der Kunsthandlung Brakl anzuschauen, und kauft einige Werke, unter anderem den eigentlich unverkäuflichen *Toten Spatz*. Aus dieser Ausstellung erwirbt auch der Verleger Reinhard Piper eine Lithographie. Der daraus entstehende Kontakt zwischen dem Maler und dem Verleger führt in den folgenden Jahren zu einer engen Zusammenarbeit.
Im April ziehen Franz Marc und Maria Franck endgültig nach Sindelsdorf. Koehler vereinbart mit ihm, daß er dem Künstler monatlich zweihundert Mark zahlt und dafür die Hälfte seiner Bilder erhält. Der Vertrag wird vorläufig für ein Jahr geschlossen, Marc ist dadurch frei von materiellen Sorgen.
Im Herbst schreibt er eine positive Kritik über die zweite Ausstellung der »Neuen Künstlervereinigung«. Dadurch kommt er auch mit diesen Künstlern in Kontakt. Hinsichtlich seiner Malerei hat er das Gefühl, auf dem richtigen Weg zu sein. Es entstehen unter anderem *Weidende Pferde I* und *Pferd in Landschaft*.

1911 An Neujahr lernt er bei Alexej von Jawlensky und Marianne von Werefkin Wassily Kandinsky und Gabriele Münter kennen. Es entsteht ein enger Kontakt, zumal Murnau und Sindelsdorf nicht sehr weit voneinander entfernt liegen.
Marc hat inzwischen die Pleinairmalerei aufgegeben. Es entstehen heute so berühmte Bilder wie *Blaues Pferd I, Der Stier, Die gelbe Kuh*.
Gegen den »Protest deutscher Künstler«, die eine Überfremdung der deutschen Kunst befürchten, initiieren Marc und Kandinsky die Schrift »Im Kampf um die Kunst«, in der sich namhafte Museumsdirektoren und Künstler äußern.
Im Mai findet seine zweite Ausstellung statt, diesmal bei Thannhauser. Anschließend fährt er mit Maria nach London, um dort zu heiraten. Diese Heirat wird in Deutschland zwar de jure nicht anerkannt, die facto leben sie aber von da an als Ehepaar zusammen.
Kandinsky stellt in einem Brief an Marc erste Überlegungen über den »Blauen Reiter« an, der Name wird im September kreiert. Im Dezember, kurz vor der Jahresausstellung, treten Kandinsky, Marc und Münter aus der »Neuen Künstlervereinigung« aus. In kurzer Zeit organisieren sie die epochemachende erste Ausstellung der »Redaktion des Blauen Reiters«.

Franz Marc, um 1912

Maria und Franz Marc, Bernhard Koehler, Heinrich Campendonk, Thomas von Hartmann; sitzend: Wassily Kandinsky, 1911

1912 Dieses Jahr begleitet Marc seine Frau über Neujahr nach Berlin. Er lernt die Maler der »Brücke« kennen und sucht sofort Graphiken von ihnen aus, um sie nach München zu schicken, wo im Februar die zweite (und letzte) »Blaue Reiter«-Ausstellung stattfindet. Bei Goltz werden diesmal jedoch ausschließlich graphische Arbeiten gezeigt. Auch Paul Klee ist vertreten. Im Mai erscheint endlich der Almanach »Der Blaue Reiter«. Herwarth Walden in Berlin stellt die Münchner Künstler als »Deutsche Expressionisten« aus.
Im Herbst fahren Marcs nach Bonn, besuchen Mackes und die Sonderbundausstellung. Gemeinsam mit August Macke fahren sie nach Paris, wo sie Robert Delaunay kennenlernen. Bei ihrer Rückkehr nach Bonn bekommt Marc die Gelegenheit, die Futuristen-Ausstellung in Köln mit zu hängen. Sowohl Delaunay als auch die italienischen Futuristen beeindrucken ihn sehr und beeinflußen im folgenden seine Arbeiten.

1913 Marcs reisen im Frühjahr nach Südtirol. Im Anschluß daran entstehen die Arbeiten *Das arme Land Tirol* und *Tirol*. Im Juni werden die beiden endlich nach deutschem Recht getraut. In diesem Sommer malt Marc seine großen Bilder *Der Turm der Blauen Pferde*, *Tierschicksale* und andere.
Außerdem plant er, zusammen mit Wassily Kandinsky, Alfred Kubin, Paul Klee, Erich Heckel und Oskar Kokoschka eine illustrierte Bibelausgabe herauszugeben.

1914 Ende April ziehen die Marcs in ihr eigenes Haus nach Ried bei Benediktbeuren. Hier entstehen die letzten großen Gemälde, teils abstrakt, teils gegenständlich. Im August meldet sich Marc sofort als Kriegsfreiwilliger, ebenso wie Macke. Im Herbst fällt August Macke, für Marc ein schmerzlicher Verlust.

1915 Marc schreibt viele Briefe an seine Frau, die bereits 1920 veröffentlicht werden. Außerdem entstehen drei Aufsätze und Aphorismen. Da es nicht möglich ist, zu malen – nur ein kleines Skizzenbuch ist erhalten –, drückt er sich schriftlich aus.

1916 Im Februar kommt die Hoffnung auf, daß Marc, dessen Einstellung zum Krieg sich inzwischen stark gewandelt hat, frühzeitig entlassen wird. Am 4. März trifft ihn ein Granatdoppelschuß bei einem Kundschaftsgang. 1917 wird sein Leichnam nach Kochel am See überführt, wo er seither begraben liegt.

Maria und Franz Marc, Fronturlaub 1915

Anmerkungen

1 21. Juni 1900; zit. nach: Franz Marc 1880-1916. Ausstellungskatalog der Städtischen Galerie im Lenbachhaus. München, 1980 (im folgenden zit. als: Kat. Lenbachhaus), S. 15
2 20. 10. 1905 an Marie Schnür; zit. nach: Klaus Lankheit: Franz Marc, Sein Leben und seine Kunst. Köln 1976 (im folgenden zit. als: Lankheit 1976), S. 28
3 zit. nach: Kat. Lenbachhaus, S. 28
4 zit. nach: Kat. Lenbachhaus, S. 22
5 August Macke/Franz Marc: Briefwechsel. Herausgegeben von Wolfgang Macke, Köln 1964 (im folgenden zit. als: Macke/Marc), S. 12
6 Macke/Marc, S. 14
7 Klaus Lankheit (Herausgeber): Franz Marc. Schriften. Köln 1978 (im folgenden zit. als: Schriften), S. 126ff
8 zit. nach: Kat. Lenbachhaus, S. 29
9 Macke/Marc, S. 25; die weiteren Zitate ebenda S. 25-48
10 zit. nach: Schriften, S. 129
11 zit. nach: Schriften, S. 98
12 zit. nach: Schriften, S. 99f.
13 zit, nach: Schriften, S. 112f.
14 Franz Marc: Briefe aus dem Feld. Herausgegeben von Klaus Lankheit und Uwe Steffen, München 1986 (im folgenden zit. als: Briefe aus dem Feld), S. 64f.
15 Lankheit 1976, S. 120
16 zit. nach: Schriften, S. 101
17 Briefe aus dem Feld, S. 65
18 W. Kandinsky: »Der Blaue Reiter« (Rückblick). 1930, in: Paul Vogt: Der Blaue Reiter. Köln 1977
19 Der Blaue Reiter. Herausgegeben von Wassily Kandinsky und Franz Marc. Dokumentarische Neuausgabe von Klaus Lankheit. München/Zürich 1984 (im folgenden zit. als: Der Blaue Reiter), S. 180
20 Der Blaue Reiter, S. 31
21 Der Blaue Reiter, S. 35f.
22 Schriften, S. 102
23 zit. nach: Expressionisten. Die Avantgarde in Deutschland 1905-1920. Ausstellungskatalog der Nationalgalerie in Berlin (DDR), 1986, S. 109
24 Schriften, S. 109
25 Franz Marc, Else Lasker-Schüler: »Der Blaue Reiter präsentiert Eurer Hoheit sein Blaues Pferd«. Karten und Briefe. Herausgegeben und kommentiert von Peter-Klaus Schuster. München 1987
26 Briefe aus dem Feld, S. 50
27 zit. nach: Lankheit 1976, S. 111
28 Macke/Marc, S. 179
29 Kat. Lenbachhaus, S. 43
30 Kat. Lenbachhaus, S. 179
31 Macke/Marc, S. 188
32 Schriften, S. 161
33 Schriften, S. 156
34 Briefe aus dem Feld, S. 128
35 Briefe aus dem Feld, S. 150f.
36 zit. nach: Lankheit 1976, S. 157

Der Verlag dankt den Museen, Sammlern und Archiven für die Überlassung von Bildvorlagen.
Die Abbildungen auf den folgenden Seiten wurden von den in der jeweiligen Bildlegende genannten Besitzern zur Verfügung gestellt: 1, 2, 6, 7, 10, 11, 17, 18, 19, 22, 24, 25, 27, 34/35, 36, 38, 41, 43, 48, 51, 57, 58, 60/61, 62, 64, 66, 70, 74, 75, 79, 81, 82, 84, 85, 90, 92, 93. Weitere Bildquellen: Nürnberg, Germanisches Nationalmuseum: 95 oben und rechts unten, Umschlagrückseite; Artothek, Peissenberg: 9, 12, 13, 14, 15, 16, 21, 29, 31, 32, 33, 39, 42, 45 (Aufnahme eines Hanfstaengl-Lichtdrucks des im Krieg verschollenen Originals), 52, 63, 65, 68, 69, 72, 73, 77, 80, 83, 88/89, 91, 94; Colorphoto Hans Hinz: 71; Archiv des Verlages: 30, 37, 40, 44, 46, 47, 49, 50, 53, 54, 78, 86, 87, 95 links unten. Die Abbildung S. 56 wurde mit freundlicher Genehmigung der Galerie im Lenbachhaus, München, aus dem Katalog »Franz Marc 1880–1916«, München 1980, entnommen.

In dieser Reihe: